This item must be returned or renewed by the
below. The loan period may be shorten
another reader.   A fine will be d

DATE OF

Mechanisms and Metal Involvement in Neurodegenerative Diseases

RSC Metallobiology Series

Editor-in-Chief:
Professor C. David Garner, *University of Nottingham, UK*

Series Editors:
Professor Hongzhe Sun, *University of Hong Kong, China*
Professor Anthony Wedd, *University of Melbourne, Australia*
Professor Barry P. Rosen, *Florida International University, USA*

Titles in the Series:
1: Mechanisms and Metal Involvement in Neurodegenerative Diseases

How to obtain future titles on publication:
A standing order plan is available for this series. A standing order will bring
delivery of each new volume immediately on publication.

For further information please contact:
Book Sales Department, Royal Society of Chemistry, Thomas Graham House,
Science Park, Milton Road, Cambridge, CB4 0WF, UK
Telephone: +44 (0)1223 420066, Fax: +44 (0)1223 420247
Email: booksales@rsc.org
Visit our website at www.rsc.org/books

Mechanisms and Metal Involvement in Neurodegenerative Diseases

Edited by

Roberta Ward and Robert Crichton
Catholique University of Louvain, Belgium
Email: roberta.ward@uclouvain.be; robert.crichton@uclouvain.be

David Dexter
Imperial College London, UK
Email: d.dexter@imperial.ac.uk

RSCPublishing

RSC Metallobiology Series No. 1

ISBN: 978-1-84973-588-9
ISSN: 2045-547X

A catalogue record for this book is available from the British Library

Published by The Royal Society of Chemistry,
Thomas Graham House, Science Park, Milton Road,
Cambridge CB4 0WF, UK

Registered Charity Number 207890

For further information see our web site at www.rsc.org

Contents

RSC Metallobiology Series No. 1
Mechanisms and Metal Involvement in Neurodegenerative Diseases
Edited by Roberta Ward, David Dexter and Robert Crichton
© The Royal Society of Chemistry 2013
Published by the Royal Society of Chemistry, www.rsc.org

Introduction

ROBERTA WARD,[b] DAVID DEXTER*[a] AND
ROBERT CRICHTON[b]

[a] Imperial College London, UK; [b] Université Catholique de Louvain,
Belgium
*Email: d.dexter@imperial.ac.uk

1.1 Outline

The complexity of the human brain is staggering, able to coordinate the
fingers, hands and feet of an organist, all playing on three different keyboards,
and to create three-dimensional images from light falling on a two-
dimensional retina. While the brain regulates all aspects of the functions of
our bodies, in our post-genomic era we are still a long way from understanding it, despite the enormous strides made in DNA sequencing. Weighing
only about 1.4 kg (only 2% of body mass), the brain accounts for 20% of our
total oxygen consumption and 25% of our glucose utilisation. However, as
the human population lives longer and longer, hand in hand with our ever-
increasing life expectancy goes an alarming increase in the incidence of
neurodegenerative disorders, affecting both cognititive and motor function.
Neurodegenerative disorders are set to overtake cancer to become the second
most common cause of death by 2040.[1] The most common of these are
dementias, the characteristic of which is decline in cognitive faculties and
occurrence of behavioural abnormalities which interfere with the capacity of
the afflicted individual to carry out normal daily activities. It usually affects
elderly individuals and occurs in many different forms, of which the most
common is Alzheimer's disease (AD). In the *World Alzheimer Report 2009*,[1]
Alzheimer's Disease International (ADI) estimated that 36 million people

RSC Metallobiology Series No. 1
Mechanisms and Metal Involvement in Neurodegenerative Diseases
Edited by Roberta Ward, David Dexter and Robert Crichton
© The Royal Society of Chemistry 2013
Published by the Royal Society of Chemistry, www.rsc.org

worldwide are living with dementia, with numbers doubling every 20 years to reach 66 million by 2030, and 115 million by 2050. Dementia prevalence increases with age; in the USA, whereas 5.0% of those aged 71–79 years are affected, this climbs to 37.4% of those aged 90 and older.[2] Dementias are chronic, progressive, long-lasting and, so far, incurable diseases, the worldwide costs of which now represent more than 1% of global GDP. In other words, if dementia care were a country, it would represent the world's 18th largest economy.[3] Of the neurodegenerative diseases affecting motor function, Parkinson's disease is the most common, and it is estimated that 6.3 million people worldwide suffer from the disease.[4] It is the second most prevalent neurodegenerative disorder, affecting 1% of those over 60, and 4% of those over 80.

1.2   Interplay Between the Peripheral Circulation and Brain: Importance of the Blood–Brain Barrier

The brain contains about 10^{11} specialized nerve cells, called neurons, which send electrical impulses at high speeds over long distances, down their axons. Each neuron can interconnect with tens of thousands of other neurons, at junctions called synapses. The human brain contains more than 10^{14} of these synaptic connections, which forge enormously complex neural circuits, undergoing continuous remodelling. The brain also contains cells of a different type, called glial cells, which account for 90% of the brain's cells and more than half of its volume. An accumulating body of work over the last two decades[5–7] has revealed that the glial cells are important regulators of synaptic connectivity, involved in the control of synapse formation, function, plasticity and elimination, both in health and disease. Figure 1.1 presents a representation of neurons and the three main types of glial cells.

There are three types of neurons: multipolar neurons, motor neurons and sensory neurons. Some collect information about our environment (both external and internal), which they transmit to other neurons, where the data are either processed or stored, while others respond to this information to regulate the control of muscle contraction, hormone synthesis, *etc.* Sensory neurons collect all sorts of information, concerning light, smell, sound, pressure, touch, *etc.*, through specialized receptors, and transform this information into electrical signals, whereas multipolar neurons receive synaptic signals from several hundred other neurons and transmit them to many other neurons at the lateral branches of their terminals. Motor neurons transmit nerve impulses to muscle cells, and their single, often very long, axons extend from the cell body of the neuron to the effector muscle cell. They have an insulating sheath of myelin, a kind of biological insulating tape, covering all parts of the axon except for the nodes of Ranvier and the axon terminals at the neuromuscular synapse. This allows them to propagate nerve impulses at velocities of up to 100 m s^{-1}, around ten times faster than in unmyelinated nerves.

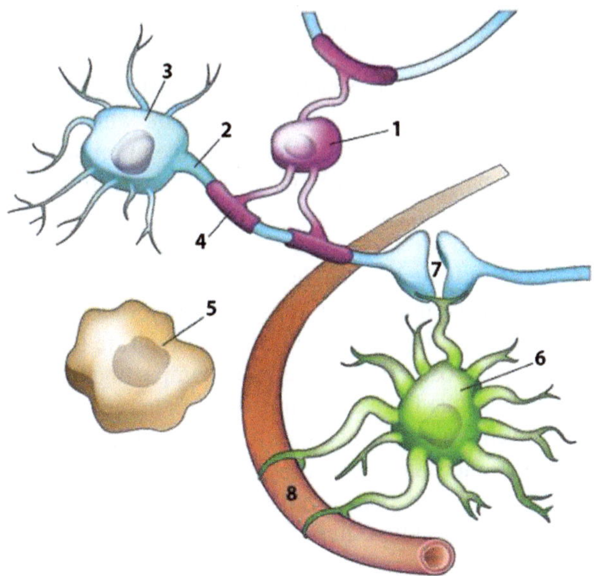

Figure 1.1 A representation of neurons and the three main types of glial cells. Oligodendrocytes (1) send projections that wrap the axons (2) – of neurons (3) – in sheaths of myelin (4), speeding signal conduction. Microglia (5) are the brain's immune cells, but they also monitor neighbouring brain cells for damage which they remove, and also have other functions. Astrocytes (6) carry out a host of activities. They can monitor levels of neuronal activity along axons at synapses (7) and, when neuronal activity is high, signal to local blood vessels (8) to dilate, increasing blood supply to the neurons. Adapted from Standmed 2009.

There are three principal types of glial cells, namely astrocytes, oligodendrocytes and microglia. Astrocytes are star-shaped glial cells which are the major cell type in the central nervous system (CNS). The network of astrocytic processes forms the infrastructure on which all other CNS cells and vessels are anchored. They have a multitude of functions, including regulation of the ionic milieu in the intercellular space, uptake and/or breakdown of some neurotransmitters, supplying nutrients to the neurons and formation of the blood–brain barrier (BBB). Astrocytes also produce and secrete substances that have a major influence on the formation and elimination of synapses.

Oligodendrocytes are involved in the electrical insulation of nerve fibres (axons), wrapping up to 150 layers of myelin sheath approximately 1 µm thick around the axons of neurons, rather like electrical insulating tape. One oligodendrocyte can extend its processes to up to 50 neuronal axons. In the peripheral nervous system the function of the oligodendrocytes is replaced by Schwann cells, which however can wrap around only one axon. High numbers of proliferating oligodentroglial progenitor cells are present to ensure remyelination can occur when necessary. New studies of human brain using magnetic resonance imaging (MRI) have demonstrated that myelinating oligodendrocytes sense electrical activity in axons, revealing that white matter changes after learning complex tasks.

Microglia are also found in the vicinity of the BBB. The microglia are the resident macrophages of the central nervous system, which can communicate with the astrocytes and neurons and with cells of the immune system by a large number of signalling pathways. They are the most susceptible sensors of brain pathology, and when they detect any signs of brain lesions or nervous system dysfunction, they undergo a complex, multistage activation process that converts them into "activated microglia". Activated microglial cells have the capacity to release a large number of substances that can act detrimentally or beneficially upon surrounding cells; they can also migrate to the site of injury, proliferate and phagocytose cells and cellular compartments. However, it has become clear that microglia have a role in the maintenance of synaptic integrity and are capable of removing defunct axon terminals, thereby helping neuronal contacts to remain intact. In the healthy CNS, microglia do not present as macrophages, indicating that their day-to-day function is different, and these specific non-macrophagic microglial functions are now beginning to be explored.

The brain is unique among all the organs of the body, hidden behind a relatively poorly permeable vascular barrier, which limits its access to plasma nutrients, such as metal ions. There are three principal barrier sites which constitute the interface between the peripheral circulation and the brain.[8] These are (Figure 1.2) the endothelium of the brain microvessels (forming the BBB proper); the epithelium of the choroid plexus, which secretes cerebrospinal fluid (CSF) into the cerebral ventricles; and the epithelium of the arachnoid mater covering the outer brain surface above the layer of subarachnoid CSF. Together the choroid plexus and the arachnoid form the blood–CSF barrier (BCSFB). The BBB is created at the level of the cerebral capillary endothelial cells, and is essentially composed (Figure 1.3) of the cerebral capillary endothelial cells, joined by tight junctions, a basal lamina, pericytes and astrocyte end-foot processes. The types of cells found at the BBB, and their associations, are illustrated in Figure 1.3. The endothelial cells form tight junctions which seal the paracellular pathway between the cells, such that substances which enter the brain must use dedicated endothelial cell transport systems. Pericytes, the connective tissue cells which occur around small blood vessels, are distributed along the length of the cerebral capillaries, partially surrounding the endothelium. Both the cerebral endothelial cells and the pericytes are enclosed by the local basement membrane, forming a distinct perivascular extracellular matrix (basal lamina 1, BL1), different from the extracellular matrix of the astroglial end-feet bounding the brain parenchyma (BL2). Foot processes from astrocytes form a complex network surrounding the capillaries.

In brain endothelium, adsorptive and receptor-mediated transcytosis allow restricted and regulated entry of certain large molecules that have particular growth factor and signalling roles within the CNS. Once the BBB is crossed, diffusion distances for solutes to neurons and glial cells are short. Unlike other

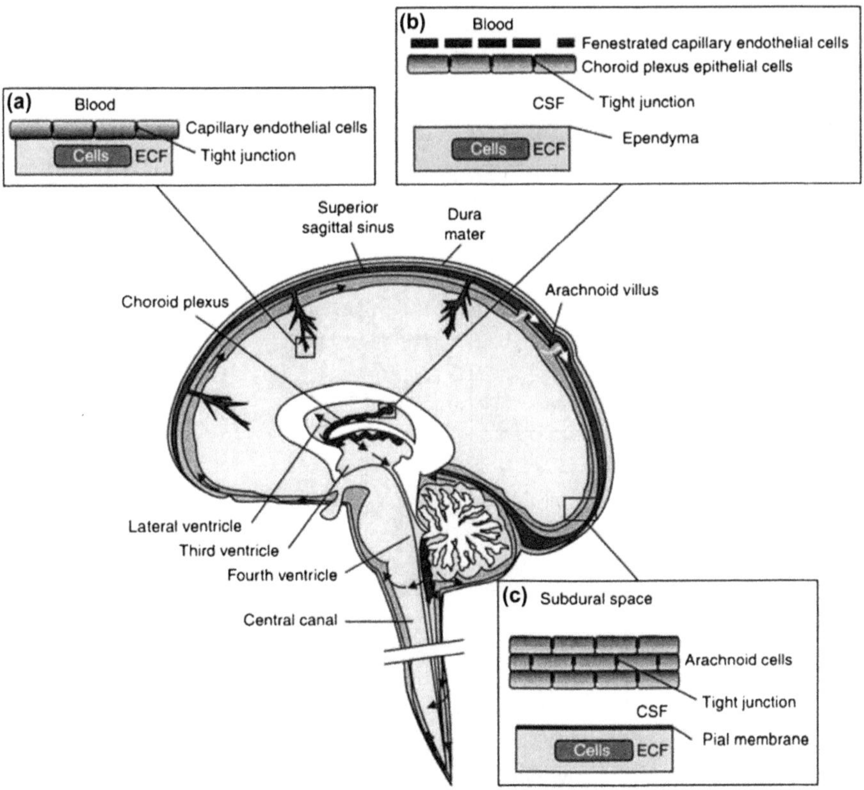

Figure 1.2 Barriers of the brain. There are three principal barrier sites between blood and brain: (a) the BBB proper; (b) the blood–CSF barrier; and (c) the arachnoid barrier.
(Reproduced from Abbott *et al.*[8] with permission from Elsevier).

blood vessel epithelia, the BBB epithelia express different receptors at the luminal membrane (facing the circulation) compared to the abluminal membrane, surrounded by astrocyte end-feet, neuronal processes and interstitial fluid.

1.3 Immune Function in Neurodegenerative Diseases

The immune system can be divided into two interactive systems: the innate and the adaptive immune systems. Innate immunity is the first line of defence, which mobilizes a rapid response to endogenous or exogenous molecules, by distinguishing self from non-self. Innate immunity is rapidly mobilized initially when endogenous ligands are released, which in turn alert the immune system and activate toll-like receptors (TLRs). In contrast, the adaptive immune system is involved in the elimination of pathogens during the later phase of infection and is elicited by B and T lymphocytes, which utilize immunoglobulins and T cell receptors, respectively, as antigen receptors to recognize "non-self" molecules.

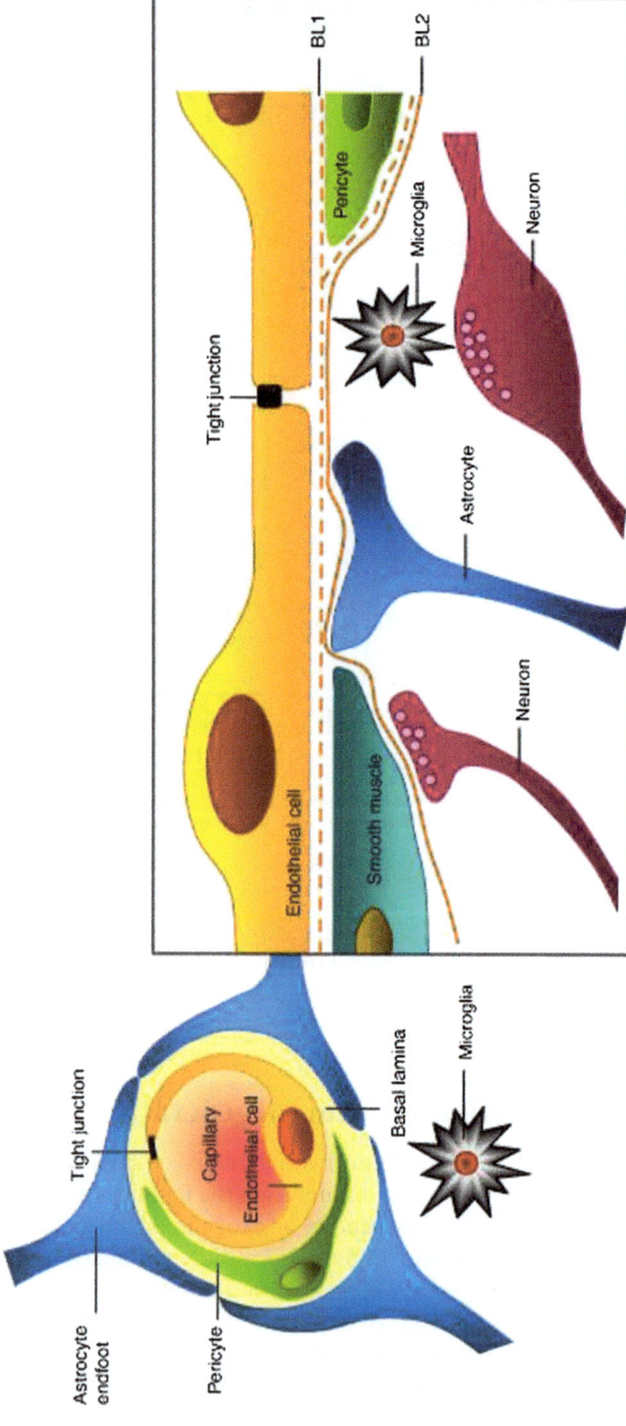

Figure 1.3 The cell associations at the BBB. (Reproduced from Abbott *et al.*[8] with permission from Elsevier).

Although the innate immune system is the dominant system of host defence in most organisms, it offers no long-term protection, whereas the adaptive immune system allows for both a stronger immune response and immunological memory.

The brain represents an immunologically privileged site which was long considered to be isolated from the central immune system owing to the presence of the BBB, lack of a draining lymphatic system to allow the uptake of potential antigens and the reputed immunoincompetence of microglia. CNS autoimmunity and neurodegeneration were presumed to be automatic consequences of the encounter of immune cells with CNS antigens. However, is now recognized[9] that the CNS is neither isolated nor passive in its interactions with the immune system and that it is involved in a constant interplay with the innate and the adaptive immune systems. Peripheral immune cells can cross the intact BBB, CNS neurons and glia actively regulate macrophage and lymphocyte responses, and microglia are most certainly immunocompetent but differ from other macrophage/dendritic cells in their ability to direct neuroprotective lymphocyte responses, playing an important role in innate immune responses of the CNS. How neuroimmune cross-talk is homeostatically maintained in neurodevelopment and adult plasticity is still not clear, although accumulating evidence suggests that neurons may also actively participate in immune responses by controlling glial cells and infiltrated T cells.[10] Many diseases of the CNS, such as stroke, multiple sclerosis (MS) and neurodegeneration, elicit a neuroinflammatory response with the objective of limiting the extent of the disease and supporting repair and regeneration.[11] However, various disease mechanisms lead to neuroinflammation contributing to the disease process itself.

Neuroinflammation occurs when there is chronic activation of the immune response in the CNS. Microglia are key players of the immune response in the CNS and the innate and adaptive immune responses triggered by microglia include the release of proinflammatory mediators. Microglia within the CNS parenchyma are activated in virtually all CNS diseases.[12] TLRs, a major family of pattern recognition receptors that mediate innate immunity but also link with the adaptive immune response, provide an important mechanism by which microglia are able to sense both pathogen- and host-derived ligands within the CNS. Upon recognition of their cognate ligands, TLR proteins initiate a signalling cascade[13] in order to turn on both common and unique pathways (Figure 1.4). All receptors of the superfamily, except TLR3, use MyD88 to initiate their signalling pathways.[14] Depending on which of the TLRs are activated, the subsequent signalling will either be MyD88 dependent or MyD88 independent, although some receptors have the ability to activate both types of pathway. Once MyD88 is bound to the TLR, IL receptor associated kinases (IRAK) 1 and 4 are phosphorylated and form oligomers with the tumour necrosis factor receptor associated factor-6 (TRAF-6). The oligomer interacts with transforming growth factor β-activated kinase (TAK1), *via* the TAK1-binding protein (TAB2), to activate IκB kinase (IκK). IκK is responsible for tagging IκB for degradation by phosphorylation and once IκB

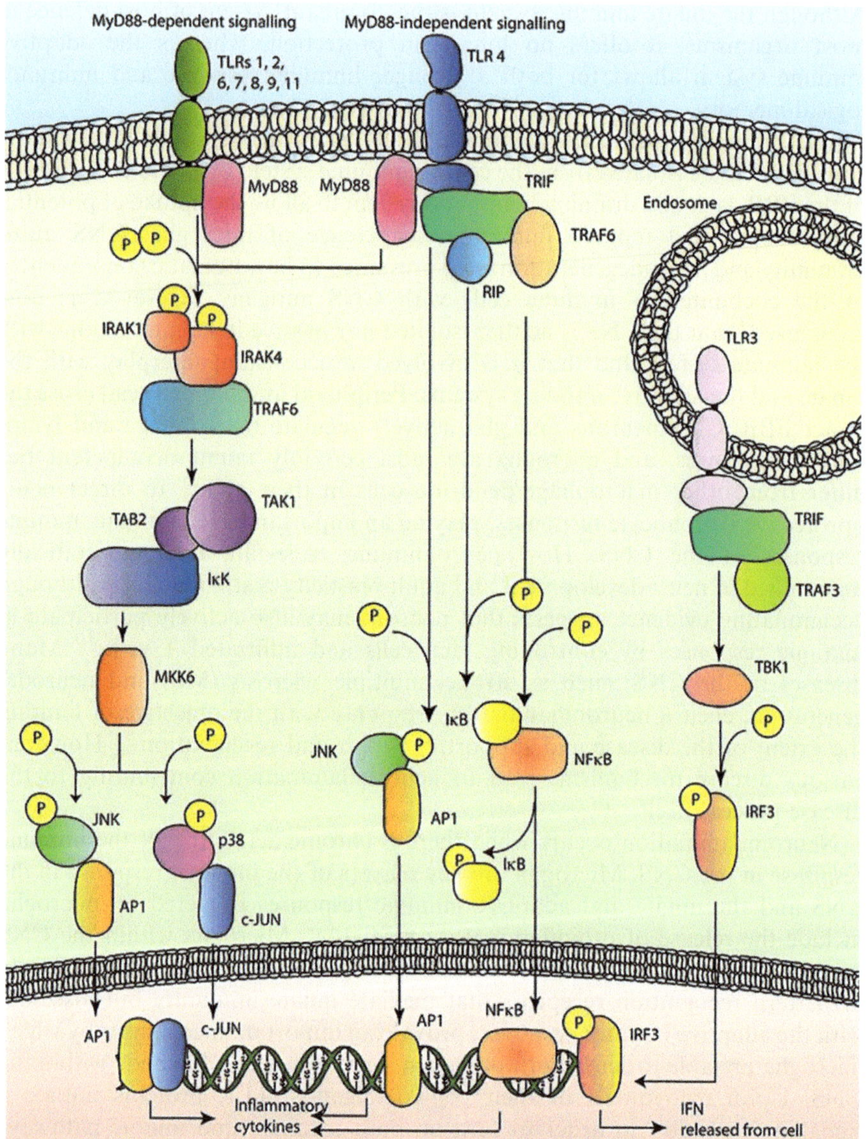

Figure 1.4 A diagram of the toll-like receptor (TLR) pathways, indicating MyD88-dependent signalling on the left, MyD88-independent signalling on the right and TLR3 only signalling on the upper right.
(Reproduced from Downes and Crack[14] with permission from Wiley).

is degraded, NFκB is free to move to the nucleus and begin transcription of various inflammation-associated genes. TAK1 is also responsible for the activation of mitogen-activated kinase kinase 6 (MKK6), which phosphorylates c-Jun N-terminal kinase (JNK) and p38. Their activation leads to nuclear translocation and transcription of inflammation-associated genes controlled by activator protein-1 (AP1) and c-JUN.

In the MyD88-independent pathway, TIR domain-containing adapter-inducing interferon-β (TRIF) binds to TRAF6, which can induce NFκB and AP1 activation, the former through interaction with IκB proteins. Receptor interacting protein 1 (RIP) can also activate NFκB by the same pathway. RIP is also essential for TLR3-mediated JNK activation. TRIF bound to TLR3 interacts with TRAF3, causing activation of TRAF family associated NFκB activator (TANK)-binding kinase (TBK1), which then phosphorylates interferon regulatory factors (IRF), releasing interferons.

Most of the TLRs and their related signalling proteins are expressed in microglia,[15] which is not surprising considering that microglia are derived from the same myeloid lineage as macrophages and dendritic cells, which represent the major sentinel cells.[16] Although there is an increasing body of evidence that TLR signalling mediates beneficial effects in the CNS, it has become clear that TLR-induced activation of microglia and the release of proinflammatory molecules are responsible for neurotoxic processes in the course of various CNS diseases. Some TLRs are expressed by astrocytes and by neurons.

The adaptive immune response involves memory B and T cells, which allows information concerning the "non-self" molecules to be inherited by all of the progeny memory B and T cells. The B cells play a key role in the generation of antibodies which circulate in the blood and lymph, generating the humoral immune response. In contrast, the T cells are intimately involved in cell-mediated immune responses, either as $CD8^+$ cytotoxic (killer) T cells or as $CD4^+$ helper T cells.

Whereas chronic neuroinflammation was previously thought to be a CD4 Th1-mediated autoimmune disease, it is now suggested that two dichotomous T cell subsets, the highly pathogenic CD4 Th17 cells and the recovery-mediating CD4 Treg cells, exert opposing effects on neurodegeneration, with the auto-aggressive Th17 cells speeding the tempo of disease while Treg cells attenuate neurodegeneration.[17]

Neuroinflammation may also be accompanied by changes in local vasculature, resulting in alterations in the permeability of the BBB, accompanied by infiltration of neutrophils, monocytes, macrophages and lymphocytes. All of this will induce microglial activation, microgliosis, and astrocyte activation, astrocytosis, with the upregulation of proinflammatory cytokines and chemokines, eventually altering synaptic plasticity and resulting in CNS dysfunction.

There are a number of anti-inflammatory systems which may play a role in regulating microglia activation. They include the highly glycosylated protein CD200 and its receptor CD200R. Binding of CD200 to CD200R helps to maintain microglia in their resting state,[18] and in both PD and AD brain, loss of CD200 in parallel with loss of neurons induces an accelerated microglia response.[19] Vitamin D3 plays a central role in immunity; its deficiency may be associated with increased prevalence of CNS diseases,[20] and upon activation by vitamin D the vitamin D3 receptor can initiate the transformation of the activated T cell either to a killer cell or to a T helper cell. The peroxisome proliferator-activated receptors, members of a superfamily of nuclear hormone receptors, play a key role in the regulation of immune and inflammatory responses,[21] downgrading the activation of microglia. Finally, the activation of the hypothalamic–pituitary–adrenal axis *via* the glucocorticoid receptor can activate gene transcription of anti-inflammatory agents and may decrease the translocation of NFκB and AP-1 to the nucleus.[21]

1.4 The Role of Metals in the Brain

A number of important biological functions in the brain require metal ions.[22] These include the fast transmission of electrical impulses between neurons and along their axons to muscles and endocrine tissues, the maintainence of ionic gradients and the synthesis of neurotransmitters. They include potassium, sodium, calcium and zinc, together with the redox-active iron and copper.

The distribution of the alkali metal ions Na^+ and K^+ in most mammalian cells is such that intracellular K^+ is high, whereas intracellular Na^+ is low. This concentration differential is required for a number of major biological processes, such as cellular osmotic balance, signal transduction and neurotransmission. It is maintained by $(Na^+\text{-}K^+)$-ATPase, which belongs to the family of P-type ATPases. The overall catalysed reaction is:

$$3Na^+(in) + 2K^+(out) + ATP + H_2O \rightleftharpoons 3Na^+(out) + 2K^+(in) + ADP + P_i$$

$$(1.1)$$

This results in the extrusion of three positive charges for every two which enter the cell, resulting in a transmembrane potential of 50–70 mV, and this has enormous physiological significance. More than one-third of the ATP utilized by resting mammalian cells is used to maintain the intracellular $Na^+\text{-}K^+$ gradient (in nerve cells this can rise up to 70%), which controls cell volume, allows neurons and muscle cells to be electrically excitable, and also drives the active transport of sugars and of amino acids.

The opening and closing of gated ion channels generates electrochemical gradients across the plasma membranes of neurons, which allow the transmission of nervous impulses not only within the brain but also the transmission of signals from the brain to other parts of the body. These nerve impulses are constituted by a wave of transient depolarization/repolarization of membranes which traverses the nerve cell, called an action potential (Figure 1.5a). The action potential results from a rapid and transient increase in Na^+ permeability, followed by a more prolonged increase in K^+ permeability (Figure 1.5b). The opening of the Na^+ channels results in an influx of 3.7 pmol cm^{-2} of Na^+, offset by a more prolonged influx of 4.3 pmol cm^{-2} of K^+ which begins a fraction of a millisecond later. The outcome is that in the first ~ 0.5 ms the membrane potential increases from the resting potential of around –60 mV to about +30 mV. The Na^+ channels now become refractory and no more Na^+

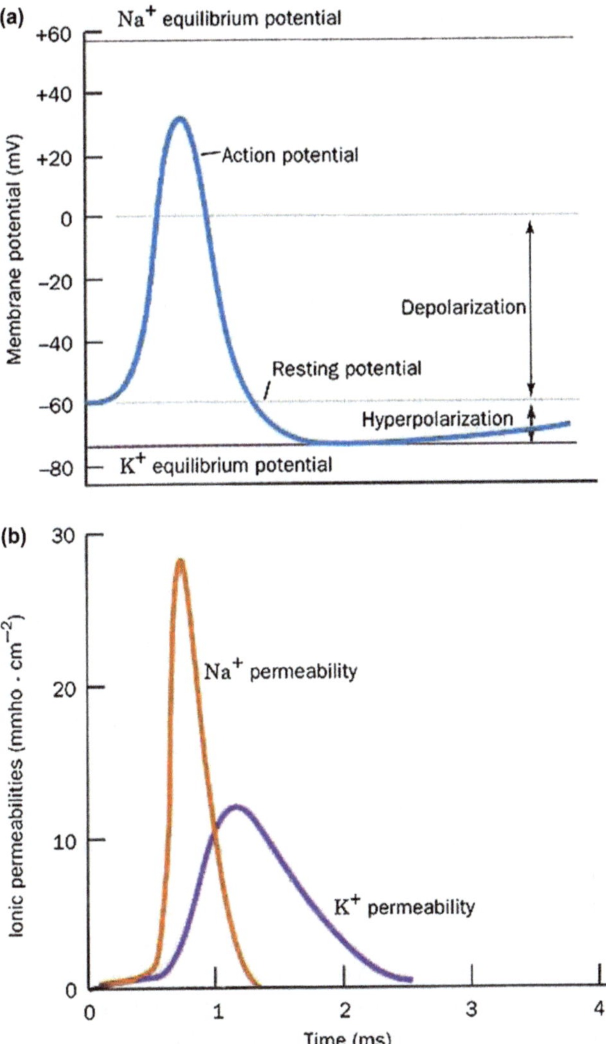

Figure 1.5   The time course of an action potential. (a) The axon membrane undergoes rapid depolarization followed by a nearly as rapid hyperpolarization and then a slow recovery to its resting potential. (b) The depolarization is caused by a transient increase in Na$^+$ permeability (conductance), whereas the hyperpolarization results from a more prolonged increase in K$^+$ permeability that begins a fraction of a millisecond later. (Reproduced from Voet and Voet[74] with permission from Wiley).

enters the cell, while K$^+$ channels remain open, causing a rapid repolarization, which allows the membrane potential to overshoot the resting potential (hyperpolarization) before the K$^+$ channels also inactivate, allowing the membrane to recover to its initial value. The voltage-gated Na$^+$ and K$^+$ ion channels across the axonal membranes create the action potentials (essentially electrochemical gradients) which allow information transfer and regulate cellular function.

The way proteins function is determined by their shape and their charge, and the binding of Ca^{2+} to proteins, just like the phosphorylation of the hydroxyl groups of Ser, Thr or Tyr residues by protein kinases (which represent about 2% of eukaryotic genomes), can trigger changes in both shape and charge. This ability of both Ca^{2+} and phosphoryl groups to alter local electrostatic fields and protein conformations are the two universal tools of signal transduction in biology. The cytoplasmic Ca^{2+} level in resting cells is extremely low (10^{-7} M), much lower than that in the extracellular fluid and in intracellular Ca^{2+} stores. This concentration gradient gives cells a superb opportunity to use Ca^{2+} as a trigger: the cytosolic Ca^{2+} concentration can be abruptly increased for signalling purposes by transiently opening Ca^{2+} channels, either ligand-gated channels, like the NMDA receptor activated by the glutamate agonist N-methyl-D-aspartate, or voltage-sensitive Ca^{2+} channels. In most cells, including nerve cells, fluxes of Ca^{2+} ions play an important role in signal transduction which regulate a wide range of cellular processes. The transient rise in cytosolic Ca^{2+} levels, initiated by extracellular signals, leads to the binding of Ca^{2+} by Ca^{2+}-sensor proteins, like calmodulin and synaptotagmin 1, which in turn activate a great variety of enzymes. One target protein for calmodulin in mammalian brain is calcineurin, a heterodimeric phosphatase, which seems to be involved in some way in synaptic plasticity.[23] Another is the Ca^{2+}/calmodulin-dependent protein kinase CaMKII, which plays a central role in Ca^{2+} signal transduction[24] and accounts for about 2% of total hippocampal protein and around 0.25% of total brain protein.[25] CaMKII is the most abundant protein in the postsynaptic density[26] (Figure 1.6), the region of the postsynaptic membrane physically connected to the ion channels which mediate synaptic transmission.

In presynaptic nerve terminals, neurotransmitters are packaged into small synaptic vesicles and released by Ca^{2+}-triggered exocytosis of synaptic vesicles at the presynaptic active zone. Work from many laboratories has provided overwhelming evidence that the Ca^{2+}-binding protein synaptotagmin 1, which localizes to the membranes of synaptic vesicles, and its homologues function as the primary Ca^{2+} sensor in most forms of exocytosis.[27]

Ca^{2+} is also involved in signalling from neuronal synapses to the cell nucleus, resulting in neuronal activity-dependent control of neuronal gene expression. This synapse-to-nucleus signalling plays a key role in circadian rhythms, long-term memory and neuronal survival. The processes of the astrocytes of the grey matter, which are highly ramified, are intimately associated with synapses, both structurally and functionallly. The extensive structural and functional association of perisynaptic glia with the synapse has given rise to the concept of the "tripartite synapse", in which synapses are defined as comprising the presynaptic and postsynaptic specializations of the neurons and the glial process that ensheaths them.[28] A schematic representation of a tripartite synapse is shown in Figure 1.6. The perisynaptic astrocyte processes contain transporters that take up glutamate (Glu, green circles) which has been released

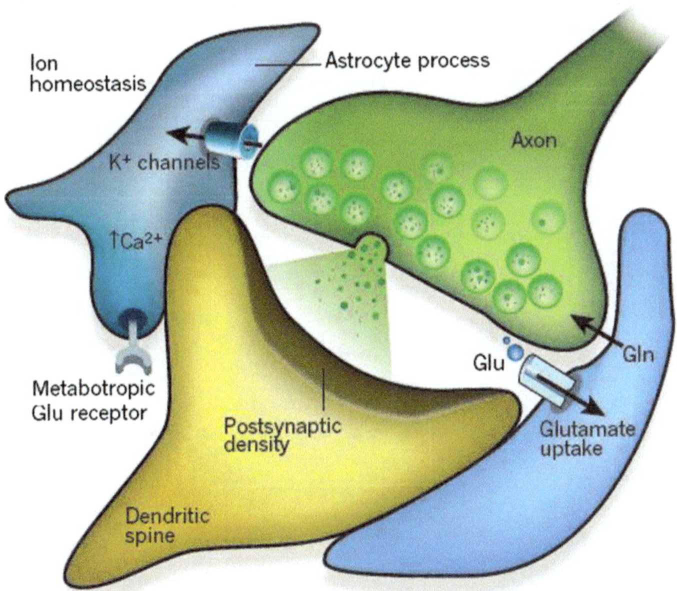

Figure 1.6   Schematic representation of a tripartite synapse in the hippocampus. The astrocyte process (*blue*) ensheaths the perisynaptic area. The axon of the neuron is shown in *green*, with the dendritic spine in *yellow*. Perisynaptic astrocyte processes contain transporters that take up glutamate (Glu, *green circles*) that has been released into the synapse and return it to neurons in the form of glutamine (Gln). Glutamate receptors on astrocytes (such as metabotropic glutamate receptors) sense synaptic glutamate release, which in turn induces a rise in Ca^{2+} concentration in the astrocytes. One of the main functions of glia at the synapse is to maintain ion homeostasis, for example regulating extracellular K^+ concentrations and pH.
(Reproduced from Eroglu and Barres[28] with permission from Macmillan).

into the synapse and return it to the neurons in the form of glutamine (Gln). Glutamate receptors on astrocytes (such as metabotropic glutamate receptors) sense synaptic glutamate release, which in turn induces an increase in Ca^{2+} concentration in the astrocytes[29,30] and stimulates the release of gliotransmitters that in turn can interact with the synaptic elements.

Another metal ion that has been extensively implicated in brain function is Zn^{2+}, the second most prevalent trace metal in the body after iron.[31–33] This means that Zn^{2+} homeostasis needs to be precisely regulated, and metallothioneins regulate its sequestration and buffering, while zinc uptake and extrusion is mediated by membrane-associated zinc transporters (Figure 1.7). Human metallothioneins (MTs) consist of two domains and bind up to seven zinc atoms, with three Zn^{2+} bound to nine cysteines in one domain and four Zn^{2+} bound to eleven cysteines in the other (Figure 1.7). In the CNS,[34] MTs show a diverse pattern of expression, with MT-1 and MT-2 mainly expressed in astrocytes and spinal glia but largely absent from neurons, whereas MT-3 is

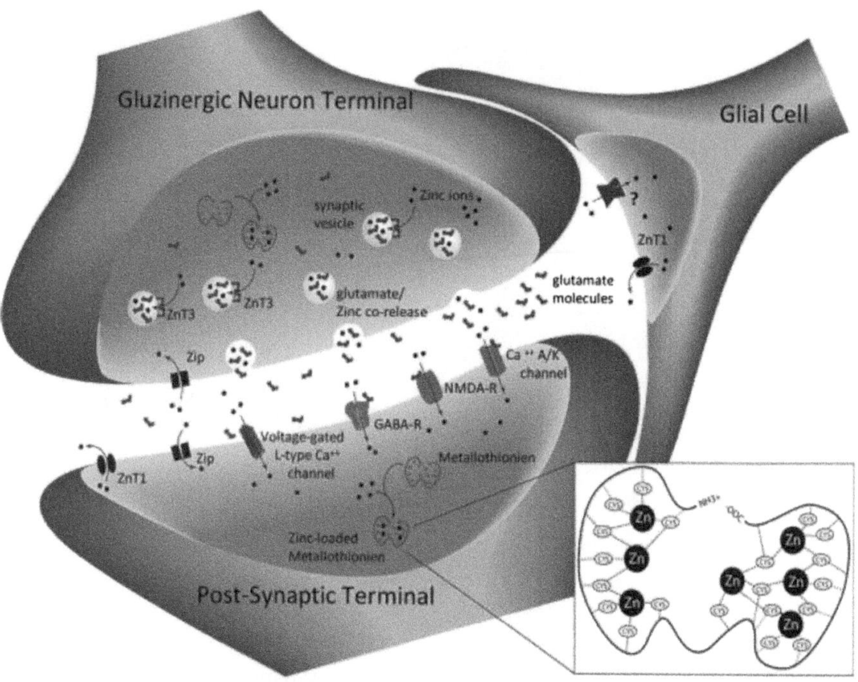

Figure 1.7 Zinc trafficking at the gluzinergic synapse. Zinc enters vesicles of gluzinergic terminals *via* the zinc transporter (ZnT3) and is stored with glutamate. During normal stimulation, zinc is released along with glutamate into the synaptic cleft where it can then act on postsynaptic channel proteins such as GABA receptors, NMDA receptors, voltage-gated channels or a number of other ion channels to alter their activity, many of which have not been well defined, *e.g.* the unknown channel illustrated on the glial cell membrane (question mark). MTs are primary intracellular zinc-buffering proteins and they regulate the availability of free zinc in presynaptic terminals and postsynaptic neurons. The metallothionein molecule (*inset*) consists of two domains, in each of which zinc is bound in a cluster. In one domain, three zinc atoms are bound to nine cysteines (cys), whereas in the other domain, four zinc atoms are bound to 11 cysteines. Each zinc atom is tetrahedrally coordinated to four thiolate bonds, with some of the thiolate ligands sharing the zinc atom.
(Reproduced from Bitanihirwe and Cunningham[31] with permission from Elsevier).

expressed exclusively in neurons and may play an important role in neuronal zinc homeostasis as it is widely distributed in the brain associated with neurons containing synaptic zinc. Mammalian zinc transporters belong to two gene families.[35] ZnT transporters facilitate zinc efflux from cells and promote zinc accumulation into intracellular vesicles, whereas the Zip family transports extracellular and vesicular zinc into the cytoplasm. There are 10 human ZnT genes, of which ZnT1 and ZnT3 have been co-localized with zinc in the synaptic vesicles. Vesicular zinc concentration is determined by the abundance of ZnT3, and ZnT3 is required for zinc transport into synaptic vesicles. ZnT1 is responsible for zinc efflux from the cell. Fourteen Zip genes have been characterized, many of which are located at the plasma membrane, although Zip7 has been identified in the Golgi apparatus.

The mammalian forebrain contains a subset of glutamatergic neurons that sequester zinc in their synaptic vesicles. Zinc-containing axon terminals are particularly abundant in the hippocampus, piriform cortex, neocortex, striatum and amygdala. Cytosolic zinc is transported into vesicles by the neuronal-specific zinc transporter ZnT3, and the vesicular glutamate transporter Vglut1 is found in the same vesicle population as ZnT3.

Zinc is released into the synaptic cleft during synaptic transmission, making zinc available for entry into neighbouring cells through gated zinc channels. Since these zinc-releasing neurons also release glutamate, the term "gluzinergic" has been proposed to describe them. While the fate of neuronally released zinc is not totally clear, it appears to modulate the overall excitability of the brain through its effect on voltage-gated calcium channels as well as on NMDA, γ-aminobutyric acid (GABA) or a number of other ion channels (Figure 1.7).[31,32] Metallothioneins are the primary intracellular buffering proteins, and they regulate the availability of free zinc in presynaptic terminals and postsynaptic neurons.

Multiple mechanisms by which such extracellular zinc could modulate fast excitatory glutamantergic receptors have been suggested. Both ionotrophic glutamate receptors and glutamate transporters are sensitive to zinc. Zinc selectively inhibits NMDA receptor-mediated responses in cultured hippocampal neurons, by producing a voltage-dependent non-competitive inhibition, resulting in a decrease in channel opening.

Glutamate receptors, which will clear glutamate from the synaptic cleft, are also modulated by zinc. Inhibition of glutamate uptake may be damaging to such activated neurons, although, since zinc will also inhibit the release of glutamate, no real change may occur within the synaptic cleft. In addition, high voltage-activated calcium channels that mediate calcium-dependent neurotransmitter release at the central synapses are also inhibited by μM concentrations of zinc. Overall it can be clearly observed that zinc could act as a critical neural messenger in both health and disease *via* its ability to regulate NMDA receptor activity. Excessive synaptic release of zinc followed by entry into vulnerable neurons contributes to severe neuronal cell death.

Genetic and nutritional studies have illustrated the essential nature of copper for normal brain function.[36] Deficiency of copper during the foetal or neonatal period will have adverse effects both on the formation and the maintenance of myelin. Excess "free" copper is, however, also dangerous, due to its capacity, like iron, to participate in redox reactions, generating toxic reactive oxygen and nitrogen species. Therefore, intracellular copper concentration must be maintained at very low levels, perhaps as low as 10^{-18} M, although how brain copper homeostasis is regulated is still not well understood. Copper serves as an essential cofactor for two key proteins involved in neurotransmitter synthesis: dopamine β-hydroxylase, which transforms dopamine to noradrenaline, and peptidyl-α-amidating monooxygenase, involved in the amidation of neuropeptides. Mitochondrial function, essential for brain function, and protection against reactive oxygen species in the cytosol rely on the activities of cytochrome C oxidase and Cu/Zn superoxide dismutase, respectively. Ceruloplasmin is also important for cellular iron mobilization (in aceruloplasminaemia, iron accumulates in brain).

The major route of copper entry into neuronal cells (Figure 1.8) is *via* the Cu^+ transporter Ctr1. Copper transport to various cuproenzymes from Ctr1 is mediated *via* metallochaperone pathways.[37] Interestingly, the amyloid precursor protein possesses an N-terminal copper binding domain which could reduce Cu^{2+} to Cu^+. The CSF contains non-ceruloplasmin bound copper, although the ligand to which it is bound is not yet identified. The copper

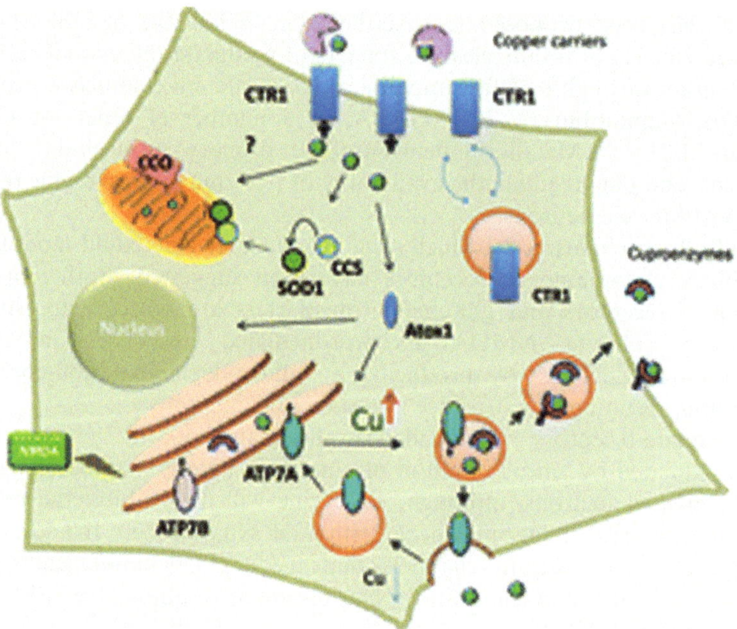

Figure 1.8 Copper distribution in a generalized cell in the CNS. In extracellular fluids, copper (*green balls*) is bound to either specific copper carriers (exchangeable copper) or to enzymes that use copper as a cofactor (cuproenzymes). Copper enters the cell *via* the high affinity copper transporter CTR1, located at the plasma membrane. The levels of CTR1 at the membrane can be regulated *via* recycling mechanism. Copper binds to the cytosolic copper chaperones CCS and Atox1, which facilitate copper delivery to SOD1 and the Cu-ATPases ATP7A and ATP7B, respectively. ATP7A and ATP7B transfer copper into the lumen of the trans-Golgi network (TGN) for incorporation into secreted and plasma membrane-bound cuproenzymes. When Cu is elevated or in response to other signals (such as activation of the NMDA receptor), ATP7A moves from the TGN and facilitates copper excretion. Whether or not ATP7B traffics in the CNS is presently unknown.
(Reproduced from Lutsenko *et al.*[36]).

transporters ATP7A and ATP7B use ATP hydrolysis to catalyse the transport of Cu^+ across membranes. It is thought that such copper is subsequently transported into intracellular vesicles, which then fuse with the plasma membrane and release the copper from the cell. Copper export can be stimulated in response to Ca^{2+} channel activation. ATP7A expression in mouse brain in early postnatal development is in the hippocampus, olfactory bulb, cerebellum and choroid plexus. This alters with ageing, with the highest ATP7A expression found in CA2 hippocampal pyramid cells, cerebellar Purkinje neurons and choroids plexus. Copper can be released from synaptic vesicles into the synaptic cleft of glutamatergic synapses in the cortex and hippocampus following depolarization. In cultured hippocampal neurons *in vitro*, the efflux of copper involves copper-independent trafficking of ATP7A to neuronal processes *via* activation of NMDA receptors.

Since iron is involved in many CNS processes[38] that could affect infant behaviour and development, iron deficiency has adverse effects on brain development, both pre- and post-natal. In contrast, an inevitable consequence of ageing is an elevation of brain iron in specific brain regions localized within H- and L-ferritin and neuromelanin, with no apparent adverse effect. However, as we will see in subsequent chapters, ill-placed excessive amounts of iron in specific intracellular compartments, or in specific regions of the brain, will lead to neurodegenerative diseases.

Iron is critical for a great number of cellular processes, and in the CNS iron is required as a co-factor for oxidative phosphorylation, nitric oxide metabolism and oxygen transport, as well as for several neuronal specific functions such as dopaminergic neurotransmitter synthesis (tryptophan hydroxylase for serotonin and tyrosine hydroxylase for noradrenaline and dopamine) and myelination of axons. The brain iron content is less than 2% of total body iron content. Nonetheless, while the iron content within different brain regions varies greatly, significantly higher iron concentrations are found in some brain regions than in the liver (Figure 1.9).[39] Regions of the brain associated with

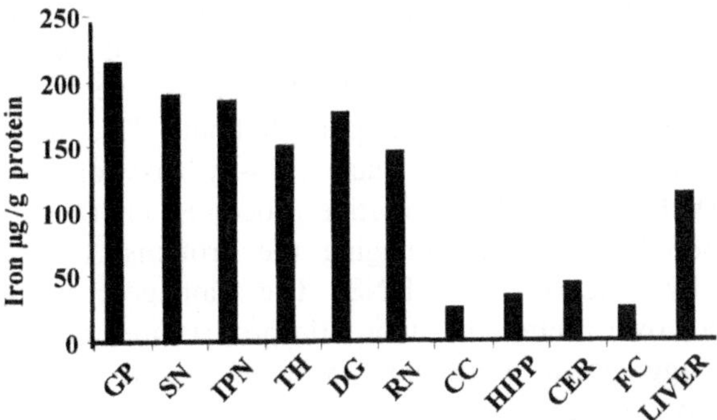

Figure 1.9   Distribution of iron in human brain. GP, globus pallidus; SN, substantia nigra; IPN, interpeduncular nucleus; TH, thalamus; DG, dentate gyrus; RN red nucleus; CC, cerebral cortex; HIPP, hippocampus; CER, cerebellum; FC, frontal cortex.
(Reproduced from Götz *et al.*[39] with permission from Wiley).

motor functions tend to have more iron than non-motor-related regions. The mechanism of transport of iron across the BBB remains conjectural;[40] however, transferrin synthesized by the oligodentrites in the brain will bind the majority of the iron that traverses the BBB after its oxidation, possibly by a glycophosphoinositide-linked ceruloplasmin found in astrocytic foot processes that surround brain endothelial cells. Neurons acquire iron from diferric transferrin. However, the source of iron within microglia cells is unclear. Ferritins can store excess iron and release it when required for cellular processes. In addition, in brain, neuromelanin (NM), an organic polymer consisting of dihydroxyindole and benzothiazine units, which are products of dopamine metabolism, is also present. NM is able to bind a number of metals, *e.g* copper and iron.

Whereas enormous advances have been made in the last 10 years in our understanding iron homeostasis within the systemic circulation,[41] how the brain regulates fluxes and storage of iron in neurons, oligodendrocytes, astrocytes and glial cells remains an enigma.

1.5 Role of Oxidative Stress in Neurodegenerative Diseases

Over the last decade, it has become more and more widely accepted that inflammation, associated with dysfunction of metal ion homeostasis (Fe, Cu, Zn) accompanied by concomitant oxidative stress, is a key factor in a large number of neurodegenerative diseases such as Alzheimer's disease (AD), Parkinson's disease (PD), Huntington's disease, amyotrophic lateral sclerosis (ALS), multiple sclerosis, Friedreich's ataxia, and others.[42] Support comes from the observation that AD, PD and many other neurodegenerative diseases are characterized by increased levels of some of these metal ions in specific regions of the brain.

The "*metal-based neurodegeneration hypothesis*" can be described[42] by the following. Redox-active metal ions (Fe, Cu), present within specific brain regions, can generate oxidative stress by production of reactive oxygen and nitrogen species (ROS, RNS), which then cause peroxidation of poly-unsaturated fatty acids in membrane phospholipids. This in turn leads to the formation of reactive aldehydes, such as 4-hydroxynonenal. The reactive aldehydes, together with other oxidative processes, interact with proteins to generate carbonyl functions, damaging the proteins, which also undergo modification by reaction with RNS. The damaged, misfolded proteins aggregate and overwhelm the ubiquitin/proteasome protein degradation system. These aggregated, ubiquinated proteins then accumulate within intracellular inclusion bodies. Such intracellular inclusion bodies are found in a great many neurodegenerative diseases (Figure 1.9).

The dioxygen molecule prefers to accept electrons one at a time, and since transition metals like iron and copper can accept and donate single electrons, they greatly facilitate its reduction. The chemistry of iron with oxygen and its two-electron reduction product, hydrogen peroxide, is outlined in

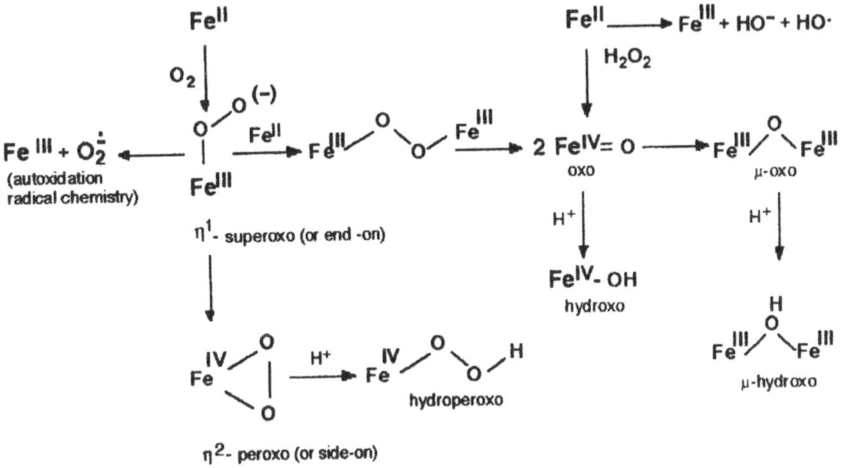

Figure 1.10   Iron–oxygen chemistry.
(Adapted from Crichton and Pierre[43]).

Figure 1.10.[43] The most reactive free radical species known, the hydroxyl radical ($^\bullet$OH), is generated by one-electron reduction of H_2O_2 in the well-known Fenton reaction[44] (eqn 1.2):

$$Fe^{2+} + H_2O_2 \rightarrow Fe^{3+} + {}^\bullet OH + OH^- \tag{1.2}$$

If superoxide could then reduce Fe^{3+} to molecular oxygen and Fe^{2+}, the sum of this reaction (eqn 1.3) plus the Fenton reaction (eqn 1.2) would result in the production of molecular oxygen plus a hydroxyl radical and a hydroxyl anion from superoxide and hydrogen peroxide (eqn 1.4) in the presence of catalytic amounts of iron: the Haber–Weiss reaction described by Haber and Weiss,[45] which is in reality thermodynamically impossible but which can function in the presence of trace amounts of iron:

$$Fe^{3+} + O_2^- \rightarrow Fe^{2+} + O_2 \tag{1.3}$$

$$O_2^- + H_2O_2 \rightarrow O_2 + {}^\bullet OH + OH^- \tag{1.4}$$

Although both ROS and RNS are involved in physiologically relevant, and important, intracellular signalling pathways, they are also associated with a number of neurodegenerative pathologies. Oxidative stress refers to a situation where elevated levels of ROS are observed, and although stimulation of ROS production by macrophages during the innate inflammatory immune response acts in a protective manner, dysregulation of ROS levels in the brain has been linked to a growing number of inflammatory and age-associated diseases.

As a necessary consequence of the production of ROS and RNS, aerobic organisms, from man to microbe, have developed an elaborate system of cytoprotection against oxidative stress. This includes both enzymes, which react directly with some of the reactive oxidants, as well as a number of

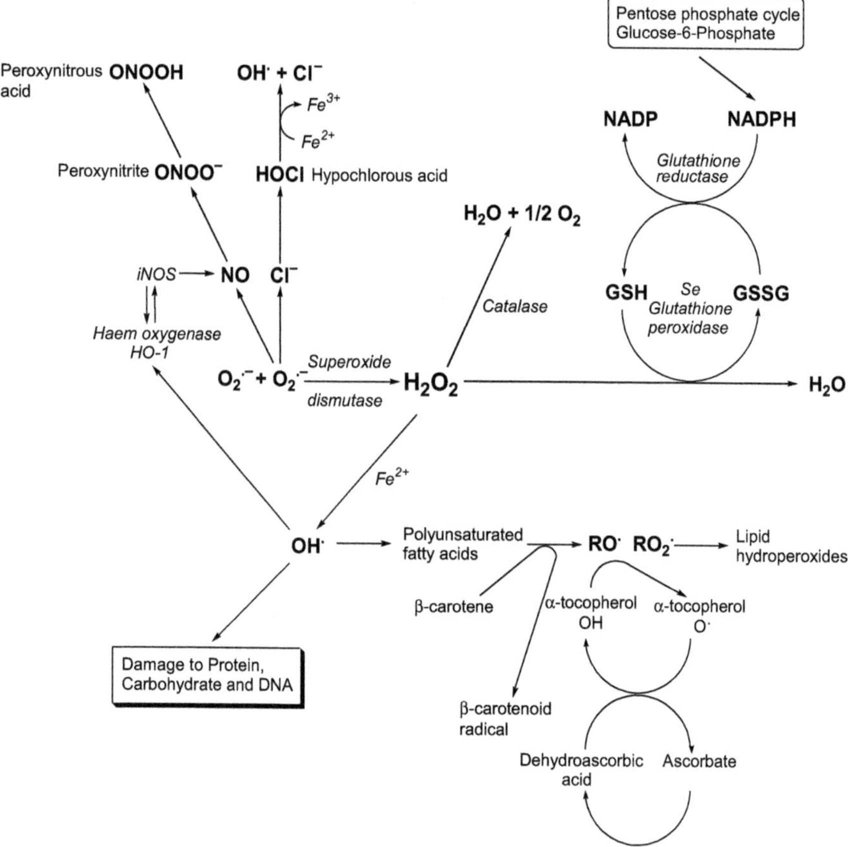

Figure 1.11   The battlefield of oxidative stress with both protagonists drawn up in horizontal and vertical lines. This could also be called the cascade of cytoprotective defence mechanisms which protect cells against the products of oxidative stress. Both defence and attack cohorts are indicated.
(Reproduced from Crichton and Ward[42] with permission from Wiley).

antioxidant molecules present within the cell. These cytoprotective enzymes and antioxidants do not act independently of one another to scavenge ROS and RNS, but function in a co-operative manner as a cascade,[46] as is illustrated in Figure 1.11.

ROS are generated by redox metals in proximity to membrane phospholipids and initiate peroxidation of polyunsaturated fatty acids (PUFAs) in the phospholipids. The lipid hydroperoxides thus generated break down through non-enzymatic Hock cleavage, to form a variety of lipid-derived α,β-unsaturated 4-hydroxyaldehydes,[47] of which the most prominent are 4-hydroxynonenal (HNE) and 4-hydroxyhex-2-enal (HHE) (Figure 1.12). The mechanisms by which these 4-hydroxyalkenals might be formed from membrane phospholipid PUFAs has been reviewed.

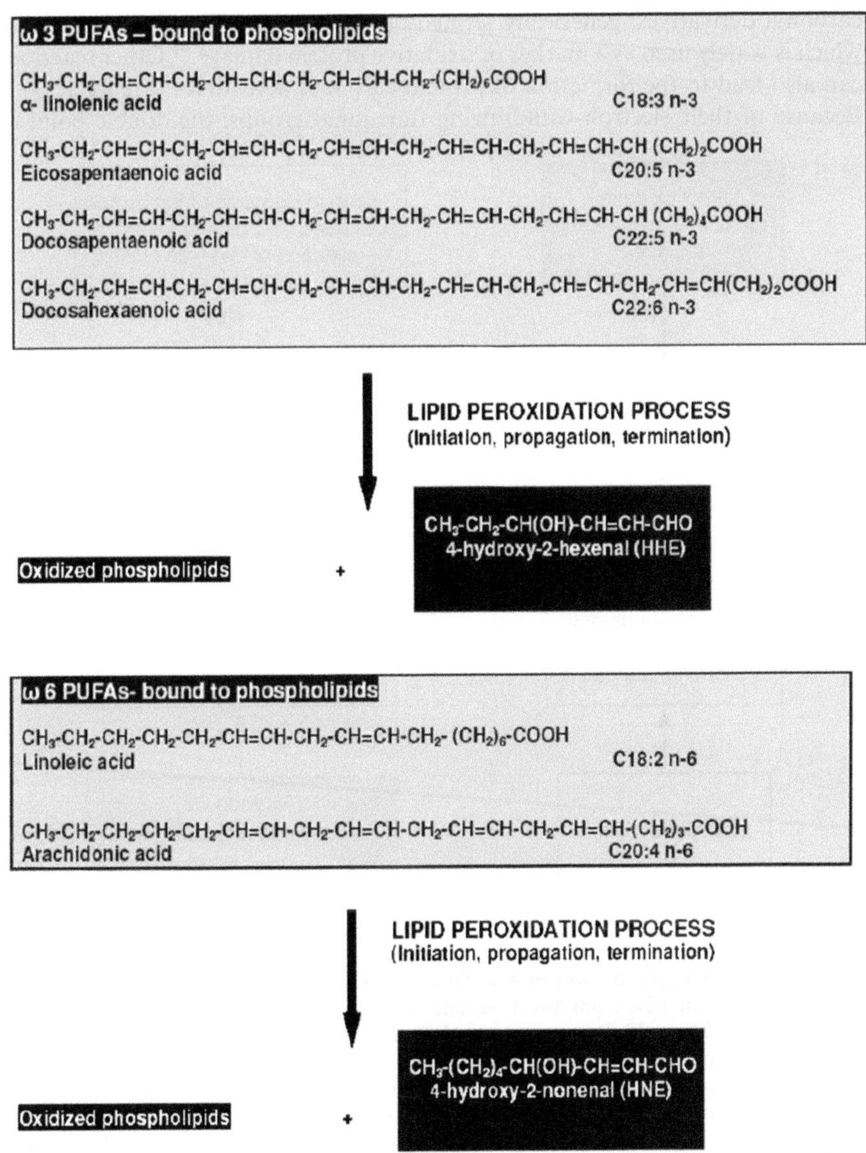

Figure 1.12   Schematic diagram of reactive hydroxyalkenals generated during lipid peroxidation of *n*–3 and *n*–6 polyunsaturated fatty acids. (Reproduced from Catala[47] with permission from Elsevier).

During oxidative stress, numerous post-translational modifications of proteins occur either from direct oxidation of amino acid residues by highly reactive oxygen species or through the conversion of lipid and carbohydrate derivatives to compounds such as reactive aldehydes which react with functional groups on proteins. A consequence is the formation of reactive protein carbonyl derivatives, generically termed "protein carbonylation", the level of which is widely used as a marker of oxidative protein damage.[48] Other reactions can also lead to the formation of protein carbonyl derivatives (Figure 1.13).[49] Because of their electron-withdrawing functional groups, the double bond of

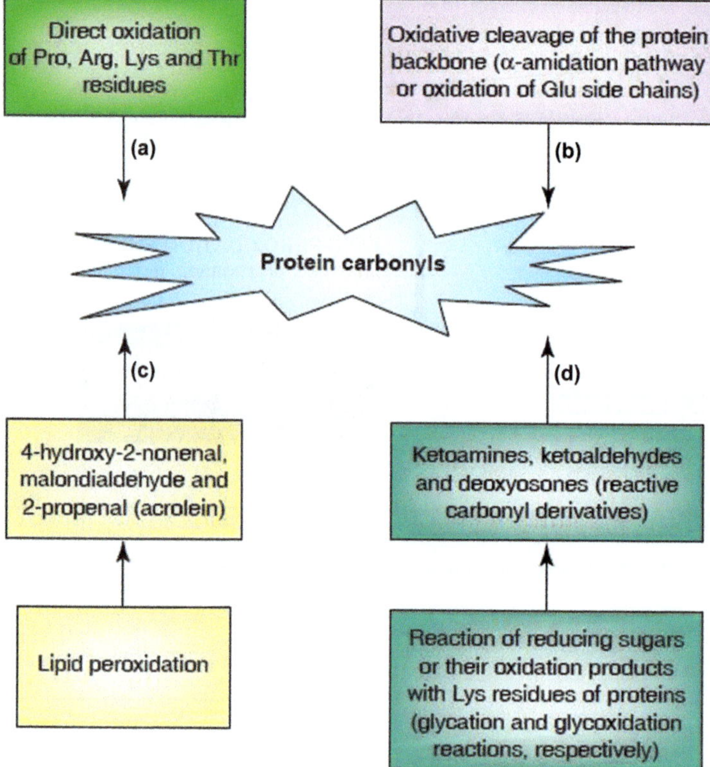

Figure 1.13   The production of protein carbonyls (aldehydes and ketones). (a) This can arise from direct oxidation of amino acid side chains (Pro, Arg, Lys, Thr). (b) Protein carbonyl derivatives can also be generated through oxidative cleavage of proteins, *via* the α-amidation pathway or through oxidation of glutamine side chains, leading to the formation of a peptide in which the N-terminal amino acid is blocked by an α-ketoacyl derivative. (c) The introduction of carbonyl groups into proteins can occur by Michael addition reactions of α,β-unsaturated aldehydes, such as 4-hydroxynon-2-enal, malondialdehyde and prop-2-enal (acrolein), derived from lipid peroxidation, with either the amino group of lysine, the imidazole moiety of histidine or the thiol group of cysteine (advanced lipoxidation end products). (d) Carbonyl groups can also be introduced into proteins by addition of reactive carbonyl derivatives (ketoamines, ketoaldehydes and deoxyosones), produced by the reaction of reducing sugars or their oxidation products, to the amino group of lysine residues, by mechanisms referred to as glycation and glycoxidation. This eventually yields advanced glycation end-products, such as carboxy-methyllysine and pentosidine.
(Reproduced from Dalle-Donne *et al.*[49] with permission from Elsevier).

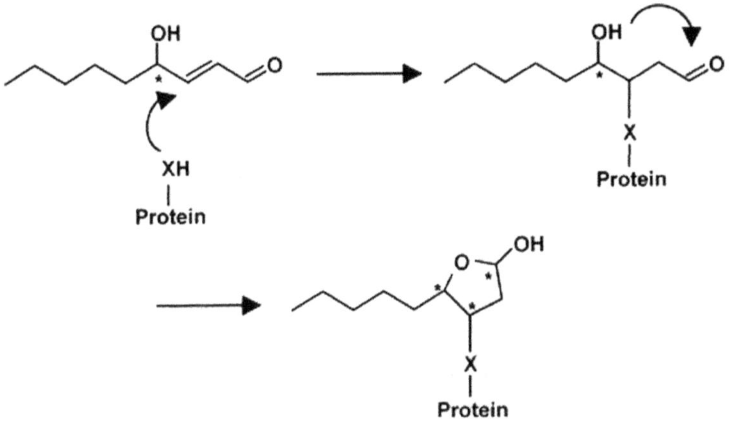

Figure 1.14 The Michael-type addition of HNE to proteins. XH: the thiol group of cysteine, the imidazole group of histidine or the amino group of lysine.

4-HNE and other α,β-unsaturated aldehydes serves as a site for Michael addition with the sulfur atom of cysteine, the imidazole nitrogen of histidine and, to a lesser extent, the amine nitrogen of lysine (Figure 1.14). After forming Michael adducts, the aldehyde moiety may in some cases undergo Schiff base formation with amines of adjacent lysines, producing intra- and/or inter-molecular cross-linking.

Of the RNS, nitric oxide itself is only a mildly reactive intermediary, but it can act as a precursor of strong oxidants under pathological conditions associated with oxidative stress. Peroxynitrite, the reaction product of nitric oxide with superoxide radicals, is one of the principal players of nitric oxide-derived toxicity that can modify thiols, oxidize methione residues and nitrate tyrosine residues in proteins. Nitration of tyrosine residues may contribute significantly to peroxynitrite toxicity, since nitration will prevent the phos-phorylation or nucleotidylation of key tyrosine residues in enzymes which are regulated by phosphorylation/adenylation, thereby seriously compromising one of the most important mechanisms of cellular regulation and signal transduction.

ROS and RNS can also readily attack DNA, generating a variety of DNA lesions. Damage to DNA bases can cause mutations and/or be cytotoxic, and repair mechanisms abound, involving about 150 repair genes in man.[50] Clearly, additional DNA damage can be induced by oxidative stress through the production of ROS in many neurodegenerative diseases.[51] More than 20 different type of base damage can be caused by ROS, of which the most common damage to purine bases is the formation of 8-oxoguanine, where the hydrogen at position 8 on guanine is replaced by an –OH group (Figure 1.15). Thymine glycol is the most common product of damage to pyrimidine bases.

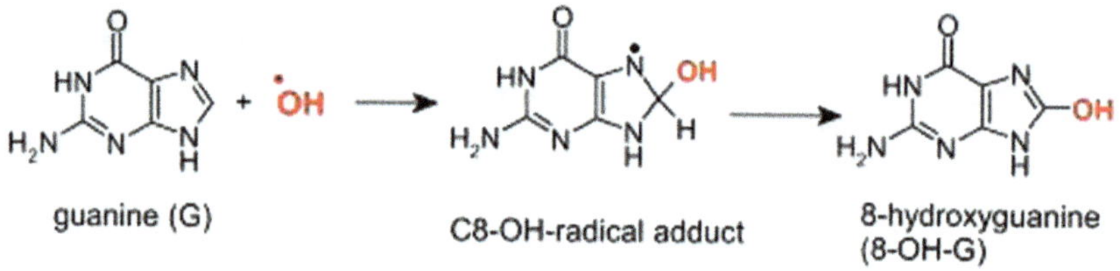

Figure 1.15 The formation of 8-hydroxyguanine by the action of the hydroxyl radical on guanine.

1.6 Importance of Protein Folding in Neurodegenerative Diseases

Most proteins adopt well-defined three-dimensional structures as a consequence of their sequence and the environment in which they fold,[52] although some examples of natively unfolded proteins are known, *e.g.* α-synuclein, tau and Aβ.[53] The ability to self-assemble and aggregate into stable, highly organized structures such as fibrils and filaments is a property shared by a large number of unrelated proteins or peptides which can be converted from their soluble forms to insoluble fibrils or plaques, known as amyloid. The term amyloid was originally used to describe protein-rich aggregates associated with a number of diseases, which were always thought to be detrimental to the host. Such amyloid fibrils have been implicated in many neurodegenerative diseases, including AD, PD and prion diseases.[54] Recently, however, it has become evident that there exist many amyloids, termed "functional amyloids," that have normal biological activities.[55] The term "amyloid" was originally used because some of the properties of amyloid fibres resemble amylose, a component of starch, including the red shift in the light absorption upon binding Congo red, and a characteristic green birefringence under polarized light. Both effects are due to the interaction of the dye molecule with the regularly spaced peptide chains, in many ways analogous to the binding of iodine to amylose. However, for protein aggregates to be properly described as amyloid, detection of the characteristic "cross-β" structures in X-ray diffraction and typical morphology in electron microscopy (EM) micrographs are generally considered as requisites.[56] The current definition of amyloid is "an unbranched protein fibre whose repeating substructure consists of β-strands that run perpendicular to the fibre axis, forming a cross-β sheet of indefinite length".[57] Thus, amyloids are composed of an ordered arrangement of many (usually thousands) copies of a peptide or protein. They are easily identified using EM as long, nonbranched filaments with diameters of 6–12 nm.[58] The repeating cross-β sheet motif gives rise to the characteristic X-ray fibre diffraction pattern shown in Figure 1.16A. The diffuse reflection along the meridian (vertical) is due to extended protein sheets running roughly parallel to the fibre, spaced 4.8 Å apart, corresponding to the inter-β strand spacing, whereas the more diffuse spacing along the equatorial reflection at ~10 Å

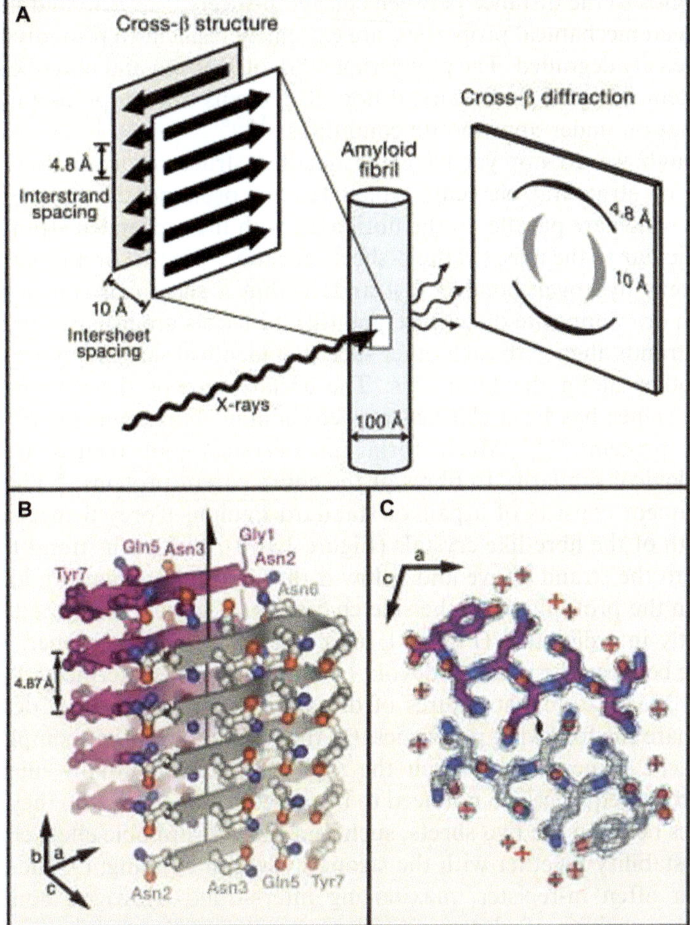

Figure 1.16   (A) The characteristic cross-β pattern observed when X-rays are directed on amyloid fibres. The diffuse reflection at a spacing of 4.8 Å along the meridian (vertical) is due to extended protein sheets running roughly parallel to the fibre, spaced 4.8 Å apart, whereas the more diffuse spacing along the equator (horzontal) shows that the extended chains are organized into sheets spaced 10 Å apart. (B) The steric-zipper structure of the sequence segment GNNQQNY from the yeast prion Sup35. Five layers of β-strands are shown of the tens of thousands in a typical fibril or microcrystal. The front sheet shows the protein backbones of the strands as *grey arrows*; the back sheet is in *purple*. Protruding from each sheet are the side chains. The *arrow* marks the fibril axis. (C) The two interdigitating β-sheets are viewed down the axis. Water molecules (shown by *red* + signs) are excluded from the tight interface between the sheets. *Red* carbonyl groups and *blue* amine groups form hydrogen bonds up and down between the layers of the sheet.
(Reproduced from Eisenberg and Jucker[61] with permission from Elsevier).

corresponds to the distance between stacked β-sheets.[59–61] Amyloid fibres have very special mechanical properties, are extremely resilient to proteolytic attack, and not easily degraded. The pioneering work of Dobson and his colleagues has clearly demonstrated that many, if not all, proteins can adopt an amyloid-like conformation under appropriate conditions.[54,62,63]

Although we do not yet have the detailed structure of amyloid fibres, a number of structural elements which feature prominently are: (i) a set of β-sheets which are parallel to the fibril axis, with their extended strands almost perpendicular to the axis; (ii) the β-sheets are either parallel or antiparallel, that is, adjacent hydrogen-bonded β-strands within a sheet can run in the same direction or in opposite directions; and (iii) the sheets are usually "in register," *i.e.* the strands align with each other such that identical side chains are on top of one another along the fibril axis. The architecture of the simplest cross-β amyloid spines has been clarified by determining short segments of amyloid-forming proteins,[64–69] which form microcrystals and related fibres with morphological similarity to fibres of the entire parent proteins.[70] The amyloid protofilament consists of a pair of standard Pauling–Corey β-sheets that run the length of the fibre-like crystals (Figure 1.16B), with each strand hydrogen-bonded to the strand above and below it through its backbone amide groups.

Within the protofilament, the side chains which project from the two sheets are tightly interdigitated (Figure 1.16C), like the teeth of a zipper. Since the interface between the sheets is devoid of water, it has been termed a "dry steric zipper". Many atomic structures of dry steric zippers have been determined, which share the following properties: (1) they usually form self-complementary amino acid sequences, in which the side chains can mutually interdigitate, although the sequences do not need to be self-complementary; (2) they have dry interfaces between the two sheets, such that the hydrophobic effect contributes to their stability together with the strong hydrogen bonding; (3) the β-strands are most often in-register, maximizing inter-strand hydrogen bonding and permitting stacking of glutamine (Gln), asparagine (Asn) and tyrosine (Tyr) residues.

The amyloid state can be formed when a segment of a protein exposes its backbone amide N–H and C=O groups, allowing them to form hydrogen bonds with other protein chains. This can occur by denaturation of normally folded proteins;[71] by overexpression of a protein overwhelming cellular chaperones and driving it into an inclusion body;[72] by cleavage of a peptide, *e.g.* Aβ from a folded protein; or by overproduction of a natively disordered protein, *e.g.* tau. In addition to exposure of backbone amide groups, the local concentration of the exposed segment must be sufficiently great to overcome the entropy that opposes formation of ordered fibres. The higher the concentration, the more aggregation is favoured. For amyloid to form, a nucleus is required to act as a template for the pattern of hydrogen bonds and steric interactions of the fibre. It has been suggested that three or four molecules must expose their amyloid-forming segments at the same time at a high enough concentration for bonding and consequent templating of the pattern to take place.[64] Thus, nucleation is a rare event, but once the nucleus is formed, single molecules can join the growing fibril one at a time, as they open to expose the proper segment and bond at the ends of fibrils. This means that amyloid fibril formation is characterized by a slow nucleation phase, followed by a more rapid growth phase.[73]

1.7 Specific Examples of Neurodegenerative Diseases

After this general introductory chapter, a series of chapters follow which illustrate a number of important neurodegenerative diseases. As far as possible, each chapter is organized in the following way. Firstly, the disease is defined and its clinical presentation is outlined. Then, if there are indications of any genetic contribution to the disease, these are described. If specific proteins are involved with the disease, the proteins and the pathology asociated with them are presented. Neuroinflammatory processes, which may imply particular metal ions associated with the disease, are highlighted, together with metabolic pathways which may be activated by these metals. If these pathways concern individual neurotransmitters, this is analysed. Finally, each chapter concludes with a detailed review of therapeutic strategies for disease treatment.

The neurodegenerative disorders which are presented are the following:

Chapter 2	Mild Cognitive Impairement	R. J. Ward
Chapter 3	Parkinson's Disease	D. T. Dexter
Chapter 4	Alzheimer's Disease	N. Hajji, C. Calvert, C.W. Ritchie, M. Sastre
Chapter 5	Friedreich's Ataxia	A. Pastore
Chapter 6	Prion Diseases	M. Rowinska-Zyrek, D. Valensin, M. Luczkowski and H. Kozlowski
Chapter 7	Multiple Sclerosis	R. R. Crichton and R. Reynolds
Chapter 8	Alcoholic Brain Damage	R. J. Ward

References

1. P. Belluck, The New York Times, April 3rd 2013.
2. B. L. Plassman, K. Langa, G. G. Fisher, S. G. Heeringa, D. R. Weir, M. B. Ofstedal, J. R. Burke, M. D. Hurd, G. G. Potter, W. L. Rodgers, D. C. Steffens, R. J. Willis and R. B. Wallace, *Neuroepidemiology*, 2007, **29**, 125.
3. *World Alzheimer Report 2011*, Alzheimer's Disease International, London, 2011.
4. R. R. Crichton and R. J. Ward, *Metal based neurodegenerative diseases: from molecular mechanisms to therapeutic perspectives*, J. Wiley and Sons, in press.
5. N. J. Allen and B. A. Barres, *Curr. Opin. Neurobiol.*, 2005, **15**, 542.
6. R. D. Fields, *Science*, 2010, **330**, 768.
7. C. Eroglu and B. A. Barres, *Nature*, 2010, **468**, 223.

8. N. J. Abbott, A. A. Patabendige, D. E. Dolman, S. R. Yusof and D. J. Begley, *Neurobiol. Dis.*, 2009, **37**, 13.
9. M. J. Carson, J. M. Doose, B. Melchior, C. D. Schmid and C. C. Ploix, *Immunol. Rev.*, 2006, **213**, 48.
10. L. Tian, L. Ma, T. Kaarela and Z. Li, *J. Neuroinflammation*, 2012, **9**, 155.
11. A. H. Jacobs, B. Tavitian and INMiND consortium, *J. Cereb. Blood Flow Metab.*, 2012, **32**, 1393.
12. G. W. Kreutzberg, *Trends Neurosci.*, 1996, **19**, 312.
13. J. A. Sloane, D. Blitz, Z. Margolin and T. Vartanian, *Neuromol. Med.*, 2010, **12**, 149.
14. C. E. Downes and P. J. Crack, *Br. J. Pharmacol.*, 2010, **160**, 1872.
15. S. Lehnardt, *Glia*, 2010, **54**, 253.
16. D. R. Greaves and S. Gordon, *Int. J. Hematol.*, 2002, **76**, 6.
17. V. Siffrin, A. U. Brandt, J. Herz and F. Zipp, *Adv. Immunol.*, 2007, **96**, 1.
18. D. Hatherley and A. N. Barclay, *Eur. J. Immunol.*, 2004, **34**, 1688.
19. R. M. Hoek, S. R. Ruuls, C. A. Murphy, G. J. Wright, R. Goddard, S. M. Zurawski, B. Blom, M. E. Homola, W. J. Streit, M. H. Brown, A. N. Barclay and J. D. Sedgwick, *Science*, 2000, **290**, 1768.
20. C. Annweiler, A. M. Schott, G. Allali, S. A. Bridenbaugh, R. W. Kressig, P. Allain, F. R. Herrmann and O. Beauchet, *Neurology*, 2010, **74**, 27.
21. C. K. Glass and S. Ogawa, *Nat. Rev. Immunol.*, 2006, **6**, 44.
22. R. R. Crichton, *Biological Inorganic Chemistry. A New Introduction to Molecular Structure and Function*, Elsevier, Oxford, 2nd edn, 2012.
23. Z. Xia and D. R. Storm, *Nat. Rev. Neurosci.*, 2005, **6**, 267.
24. A. Hudmon and H. Schulman, *Annu. Rev. Biochem.*, 2002, **71**, 473.
25. N. E. Erondu and M. B. Kennedy, *J. Neurosci.*, 1985, **5**, 3270.
26. X. B. Liu and K. D. Murray, *Epilepsia*, 2012, **53**(suppl. 1), 45.
27. Z. P. Pang and T. C. Sudhoff, *Curr. Opin. Cell Biol.*, 2010, **22**, 496.
28. C. Eroglu and B. A. Barres, *Nature*, 2010, **468**, 223.
29. J. W. Dani, A. Chernjavsky and S. J. Smith, *Neuron*, 1992, **8**, 429.
30. X. Wang, N. Lou, Q. Xu, G. F. Tian, W. G. Peng, X. Han, J. Kang, T. Takano and M. Nedergaard, *Nat. Neurosci.*, 2006, **9**, 816.
31. B. K. Bitanihirwe and M. G. Cunningham, *Synapse*, 2009, **63**, 1029.
32. P. Paoletti, A. M. Vergnano, B. Barbour and M. Casado, *Neuroscience*, 2009, **158**, 126.
33. S. L. Sensi, P. Paoletti, A. I. Bush and I. Sekler, *Nat. Rev. Neurosci.*, 2009, **10**, 780.
34. M. Vašák and G. Meloni, *J. Biol. Inorg. Chem.*, 2011, **16**, 1067.
35. D. J. Eide, *Biochim. Biophys. Acta*, 2006, **1763**, 711.
36. S. Lutsenko, A. Bhattacharjee and A. L. Hubbard, *Metallomics*, 2010, **2**, 596.
37. L. Banci, I. Bertini, K. S. McGreevy and A. Rosato, *Nat. Prod. Rep.*, 2010, **27**, 695.
38. R. R. Crichton, *Iron Metabolism: From Molecular Mechanisms to Clinical Consequences*, Wiley, Chichester, 3rd edn, 2009.

39. M. E. Götz, K. Double, M. Gerlach, M. B. Youdim and P. Riederer, *Ann. N. Y. Acad. Sci.*, 2004, **1012**, 193.

40. T. Moos, T. Rosengren Nielsen, T. Skjørringe and E. H. Morgan, *J. Neurochem.*, 2007, **103**, 1730.

41. M. W. Hentze, M. U. Muckenthaler, B. Galy and C. Camaschella, *Cell*, 2010, **142**, 24.

42. R. R. Crichton and R. J. Ward, *Metal-based Neurodegeneration: From Molecular Mechanisms to Therapeutic Strategies*, Wiley, Chichester, 2006.

43. R. R. Crichton and J. L. Pierre, *Biometals*, 2001, **14**, 99.

44. H. J. H. Fenton, *Trans. Chem. Soc.*, 1894, **65**, 899.

45. F. Haber and J. Weiss, *Proc. R. Soc. London, Ser. A*, 1934, **147**, 332.

46. R. R. Crichton, *Iron Metabolism:From Molecular Mechanisms to Clinical Consequences*, Wiley, Chichester, 2nd edn, 2001.

47. A. Catalá, *Chem. Phys. Lipids*, 2009, **157**, 1.

48. I. Dalle-Donne, G. Aldini, M. Carini, R. Colombo, R. Rossi and A. Milzani, *J. Cell Mol. Med.*, 2006, **10**, 389.

49. I. Dalle-Donne, D. Giustarini, R. Colombo, R. Rossi and A. Milzani, *Trends Mol. Med.*, 2003, **9**, 169.

50. R. D. Wood, M. Mitchell, J. Sgouros and T. Lindahl, *Science*, 2001, **291**, 1284.

51. M. L. Hegde, P. M. Hegde, K. S. Rao and S. Mitra, *J. Alzheimer's Dis.*, 2011, **24**(suppl. 2), 183.

52. C. B. Anfinsen, *Science*, 1973, **181**, 223.

53. A. L. Fink, *Curr. Opin. Struct. Biol.*, 2005, **15**, 35.

54. F. Chiti and C. M. Dobson, *Annu. Rev. Biochem.*, 2006, **75**, 333.

55. D. M. Fowler, A. V. Koulov, W. E. Balch and J. W. Kelly, *Trends Biochem. Sci.*, 2007, **32**, 217.

56. J. D. Sipe and A. S. Cohen, *J. Struct. Biol.*, 2000, **130**, 88.

57. J. Greenwald and R. Riek, *Structure*, 2010, **18**, 1244.

58. M. Sunde and C. Blake, *Adv. Protein Chem.*, 1997, **50**, 123.

59. W. T. Astbury, S. Dickinson and K. Bailey, *Biochem. J.*, 1935, **29**, 2351.

60. M. Sunde, L. C. Serpell, M. Bartlam, P. E. Fraser, M. B. Pepys and C. C. Blake, *J. Mol. Biol.*, 1997, **273**, 729.

61. D. Eisenberg and M. Jucker, *Cell*, 2012, **148**, 11880.

62. C. M. Dobson, *Trends Biochem. Sci.*, 1999, **24**, 329.

63. M. Vendruscolo, T. P. Knowles and C. M. Dobson, *Cold Spring Harb. Perspect. Biol.*, 2011, **3**; doi: 10.1101/cshperspect.a010454.

64. R. Nelson, M. R. Sawaya, M. Balbirnie, A. Ø. Madsen, C. Riekel, R. Grothe and D. Eisenberg, *Nature*, 2005, **435**, 773.

65. M. R. Sawaya, S. Sambashivan, R. Nelson, M. I. Ivanova, S. A. Sievers, M. I. Apostol, M. J. Thompson, M. Balbirnie, J. J. Wiltzius, H. T. McFarlane, A. Ø. C. Riekel and D. Eisenberg, *Nature*, 2007, **447**, 453.

66. M. I. Ivanova, S. A. Sievers, M. R. Sawaya, J. S. Wall and D. Eisenberg, *Proc. Natl. Acad. Sci. U. S. A.*, 2009, **106**, 18990.

67. J. J. Wiltzius, M. Landau, R. Nelson, M. R. Sawaya, M. I. Apostol, L. Goldschmidt, A. B. Soriaga, D. Cascio, K. Rajashankar and D. Eisenberg, *Nat. Struct. Mol. Biol.*, 2009, **16**, 973.
68. M. I. Apostol, M. R. Sawaya, D. Cascio and D. Eisenberg, *J. Biol. Chem.*, 2010, **285**, 29671.
69. S. A. Sievers, J. Karanicolas, H. W. Chang, A. Zhao, L. Jiang, O. Zirafi, J. T. Stevens, J. Munch, D. Baker and D. Eisenberg, *Nature*, 2011, **475**, 96.
70. M. Balbirnie, R. Grothe and D. S. Eisenberg, *Proc. Natl. Acad. Sci. U. S. A.*, 2001, **98**, 2375.
71. F. Chiti, P. Webster, N. Taddei, A. Clark, M. Stefani, G. Ramponi and C. M. Dobson, *Proc. Natl. Acad. Sci. U. S. A.*, 1999, **96**, 3590.
72. L. Wang, S. K. Maji, M. R. Sawaya, D. Eisenberg and R. Riek, *PLoS Biol.*, 2008, **6**, e195.
73. J. T. Jarrett and O. T. Lansbury, *Cell*, 1993, **73**, 1055.
74. D. Voet and J. G. Voet, *Biochemistry*, Wiley, Hoboken, NJ, 3rd edn, 2004.

Mild Cognitive Impairment

ROBERTA J. WARD

Université Catholique de Louvain, Belgium
Email: roberta.ward@uclouvain.be

2.1 Introduction

As the aging population throughout the world steadily increases, within the next two decades cognitive dysfunction will emerge as a major clinical and economic problem. Its diagnosis is becoming progressively easier by the use of a wide range of cognitive tests. However, once diagnosed it is apparent that therapeutic intervention is still not possible, with the underlying disease process either remaining static or progressing to Alzheimer's disease (AD).

2.2 Definition of Disease and Clinical Presentation

During normal brain aging, a number of physiological, biochemical and neuronal changes occur in most brain regions which have been associated with declining cognitive performance. Mild cognitive impairment (MCI) predominantly occurs with advancing age and is characterized by (a) a decline in cognitive abilities which are of insufficient severity to constitute dementia and (b) neuropathological changes.

2.2.1 Cognitive Aspects of MCI

There are two types of cognition: trait intelligence and age-induced changes, the former not being affected by the aging process. Attention and executive

RSC Metallobiology Series No. 1
Mechanisms and Metal Involvement in Neurodegenerative Diseases
Edited by Roberta Ward, David Dexter and Robert Crichton
© The Royal Society of Chemistry 2013
Published by the Royal Society of Chemistry, www.rsc.org

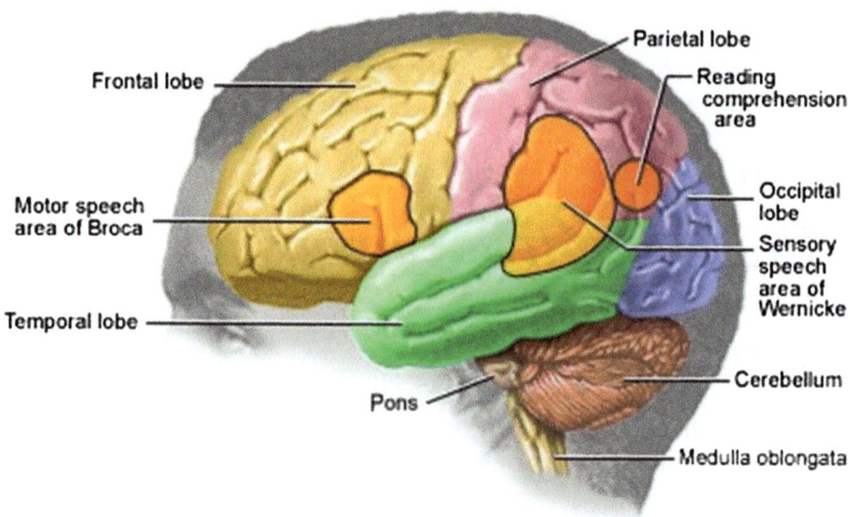

Figure 2.1   Regions of the brain implicated in MCI.

functions depend on numerous neuronal networks, which include the frontal and parietal associative cortices, the subcortical structures and their inter-connecting white matter tracts. The executive functions, which control and monitor task performance, are dependent on three fronto-subcortical circuits: the dorsolateral prefrontal cortex (working memory), the lateral orbital cortex (inhibition) and the anterior cingulate cortex (response conflict) (Figure 2.1).[1] With cognitive aging, memory deficits in healthy older adults are primarily of source memory (SM) rather than of item memory (IM). Item memory refers to remembering *what occurred* in the past, whereas source memory refers to remembering *where*, *when* and *how* something happened. Age-related impairments for SM are greater than for IM.[2] The hippocampus is the brain region critical for source memory, which will "bind" different aspects of a complex event into an integrated memory trace, thereby allowing source memory and recollection.[3]

It is estimated that in the elderly population throughout Europe and North America who are aged over 60 years, approximately 20% will exhibit mild cognitive decline.[4] Such cognitive changes are characterized by loss of memory and confusion, as well as agitation. MCI will not normally significantly interfere with daily activities. When memory loss is the predominant symptom it is called amnestic MCI. This could be an early stage of AD, although only 10–15% of such patients will progress to AD each year (Figure 2.2), while many will remain stable over time or may even ultimately show remission. When individuals with MCI show impairment of other functions, apart from memory, it is termed non-amnestic MCI or multiple domain MCI (Petersen's criteria) (Figure 2.3 and Table 2.1).[4]

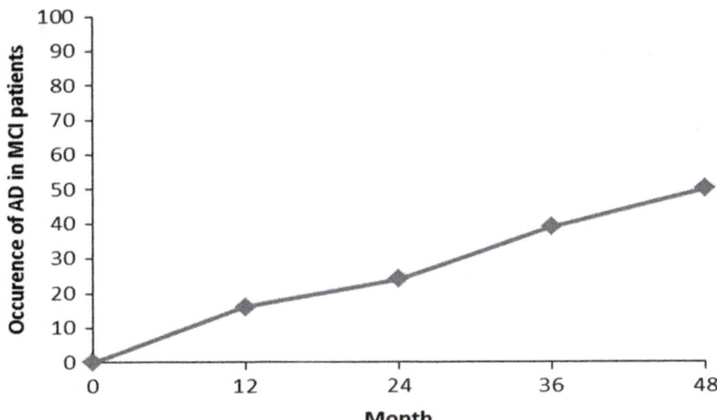

Figure 2.2 Progression of MCI to Alzheimer's disease over a four-year period.

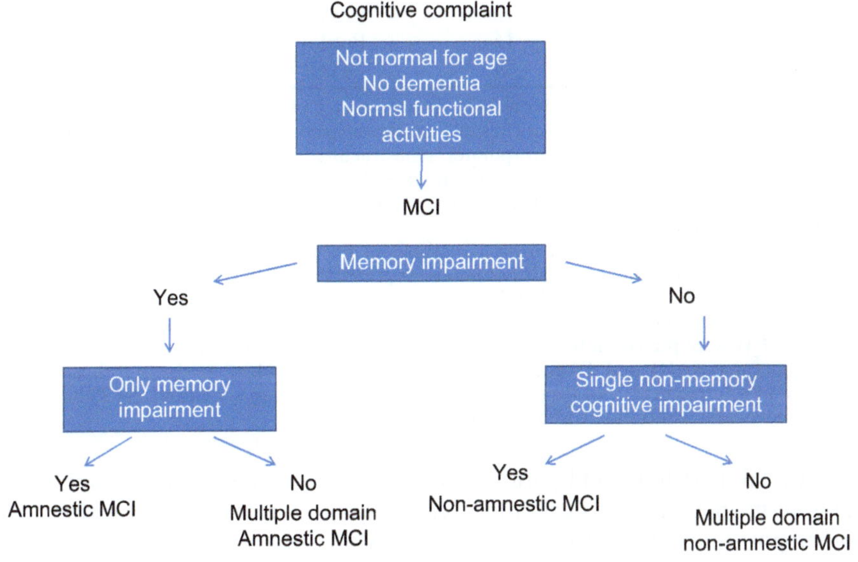

Figure 2.3 Criteria for identification of MCI.

2.2.2 Neuropathological Changes with MCI

A major problem for the identification of hallmark neuropathological changes associated with MCI post-mortem, has been the heterogeneity of the post-mortem material studied, and more importantly to understand whether such changes in the MCI brain are due to a progression to AD or not. Markesbery[5] considered that most autopsied amnestic MCI will progress to AD, and identified the presence of neurofibrillary tangles in the medial temporal lobe, *i.e.* hippocampus and adjacent anatomically related cortex (including

Table 2.1 Criteria for identification of MCI, given by the MCI Working Group of the European Consortium on Alzheimer's Disease, Brescia Meeting, Italy, June 2005.

1. Cognitive complaints coming from the patients or their families
2. The reporting of a decline in cognitive functioning relative to previous abilities during the past year by the patient or informant
3. Cognitive disorders as evidenced by clinical evaluation (impairment in memory or in another cognitive domain)
4. Absence of major repercussions on daily life (the patient may, however, report difficulties concerning complex day-to-day activities
5. Absence of dementia

Table 2.2 Clinical pathological reports of longitudinally followed subjects with MCI (adapted from [5])

	Criteria for MCI diagnosis	*Number of subjects*	*Pathology*	*Braak score*
Study 1	Petersen CDR 0.5	10	Predominant NFT in MTL	3.3
Study 2	Petersen CDR 0.5	15	Predominat NFT in MTL	2.9
Study 3	Petersen *et al.*	34	24 out of 31 AD	4.1

CDR Clinical Dementia Rating Score. The Washington University Clinical Dementia Rating (CDR) is a global scale developed to clinically denote the presence of Alzhimer type and stage its severity. The clinical protocol incorporates semistructured interviews with the patient and informant to obtain information necessary to rate the subject's cognitive performance in six domains: memory, orientation, judgment and problem solving, community affairs, home and hobbies, and personal care.
The global CDR score is derived from a synthesis of the individual ratings in each of the six domains on a 5-point ordinal scale (dichotomous for presence or absence of dementia): CDR 0 is no dementia and CDR 0.5, **1, 2,** and **3** indicate questionable, mild, moderate, and severe dementia.
Braak Score- Neurofibrillary pathology scored by Braak and Braak to follow a progressive pattern, Stage I NFTs were present in the transentorhinal area, Stage II spreading to the entohinal cortex and hippocampus. III and IV increases in the NFT were evident in the entorhinal cortex, hippocampus, adjacent inferior temporal cortex and amydala. In stages V and VI, NFTs had spread to the neocortical association cortex and other brain regions.

entorhinal, perirhinal and parahippocampal cortices, essential for establishing long-term memory for facts and events; declarative memory). Furthermore, other post-mortem studies of brain from MCI subjects have shown that the subjects with a low clinical dementia rating (CDR) score, a Braak score of between II and III (both scores defined in Table 2.2) during their latter years, showed the presence of neurofibrillary tangles in the medial temporal lobe (MTL, Table 2.2). In contrast, Mufson *et al.*[6] suggested that Aβ amyloid plaques might be a hallmark lesion of MCI, with the amyloid plaques being generated from dying neurons. Initially, Aβ plaques are present throughout the neocortex; in a second stage the Aβ plaques occur in allocortical regions; while in the third phase they are present in the basal ganglia, thalamus and hypothalamus. In a fourth stage, amyloid reaches the midbrain and medulla oblongata. Finally, at the fifth stage, Aβ is found in the pons and cerebellum. However, since Aβ plaques are present in many control brains, their usefulness as a true hallmark of MCI remains unclear. People with a CDR of 0.5 displayed an increase of diffuse plaques in the temporal cortex,[7] while non-MCI subjects, with a Braak score of II, showed no differences in the number of diffuse plaques in the neocortex or MTL. It would therefore appear that β-amyloid deposition is not a major pathological factor defining MCI. It has been suggested that Aβ mononers, dimers and higher oligomeric forms may be the toxic forms of amyloid. Aβ oligomers accumulate in the frontal cortex of MCI subjects (CDR score = 0.5) compared to age-matched controls (reviewed by Vana *et al.*[7]), with increased Aβ oligomer levels being correlated with the severity of cognitive impairment. This has led to the suggestion that Aβ monomers and oligomers underlie amyloid toxicity, causing a disruption of synaptic function in MCI, which could initiate cognitive decline. It is of interest that it has recently been suggested that Aβ is an antimicrobial peptide in the innate immune system and involved in response to clinically relevant pathogenic microorganisms.

There is substantial evidence that the structural integrity of the hippocampus declines during aging (reviews[8,9]), which is associated with deficits in source memory encoding (Figure 2.4). In addition, some studies have reported an association between changes in the prefrontal cortical and attention/executive impairment in MCI patients (reviewed[1]). A commonly observed pattern which occurs with aging is neural compensation, where a decrease in activation in the posterior cortices, (occipital cortex) is associated with an increase in anterior or frontal activity.[10] For example, functional connectivity between the hippocampus and the rest of the brain is altered with aging; the connections with posterior cortices are weaker, and connections with anterior cortices, including prefrontal regions, are stronger in older adults. Changes in grey matter within medial temporal lobe structures have been identified in some

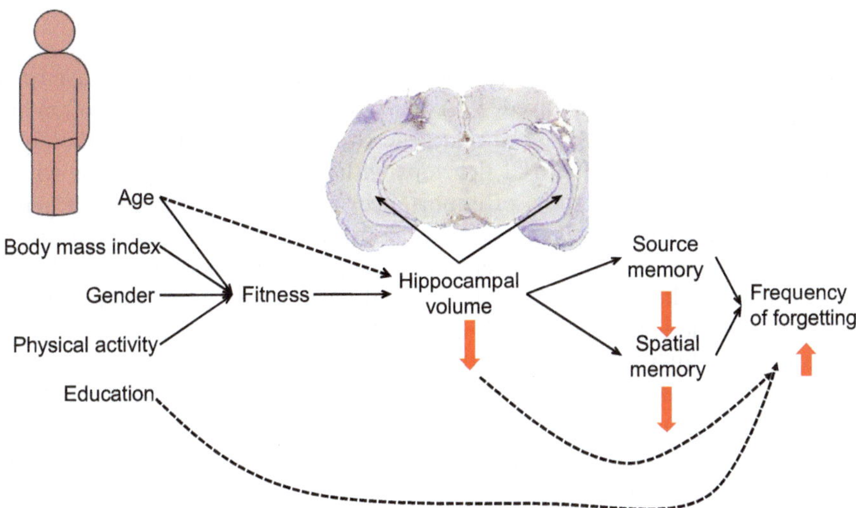

Figure 2.4 Factors affecting hippocampal volume and memory in the elderly.

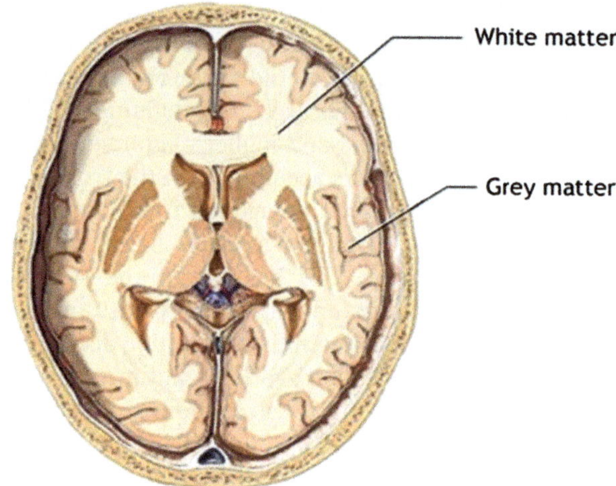

Figure 2.5   Distribution of white and grey matter in the human brain.

MCI subjects. Although white matter pathology may also be associated with age-related cognitive impairment (Figure 2.5) (and could be an early event in the development of dementia[11]), wide variations in the pattern of changes in white matter have been observed. This is possibly related to the wide heterogeneity of the MCI subjects studied, as well as the incidence of cardiovascular disease in such subjects. Therefore the importance of such changes in white matter in the brains of MCI subjects requires further investigation.

2.3   Gene Involvement and Epigenetics

MCI is a genetically complex condition but currently no major genes have been identified which may be involved. Vascular pathology and depression often are underlying factors for MCI, such that genes which influence these pathologies could be implicated. Since MCI develops with increasing aging, genes which dictate life expectancy may be of importance. The search for candidate genes for longevity, as well as genome-wide association studies, has identified a variety of genes. Fourteen independent single nucleotide polymorphisms (SNPs) that predicted risk of death, and eight SNPs that predicted event-free survival, were identified in a study by Walter *et al.*,[12] which were in or near genes that are highly expressed in the brain (HECW2, HIP1, BIN2, GRIA1), that are involved in neural development and function (KCNQ4, LMO4, GRIA1, NETO1) and, for one case, in autophagy (ATG4C). Deelan *et al.*[13] identified one major locus, rs2075650, in a genome-wide association study, which determined familial longevity.

Oxidative markers and inflammatory genes show increased expression profiles in the neocortex and cerebellum with increasing age in mice (5 months *vs.* 30 months),[14] in addition to a reduction of neurotrophic response genes. Jiang *et al.*[15] identified altered expression of 98 genes in the cortex and hypothalamus of aged mice, which included proteins involved in the oxidative stress response. Harris *et al.*[16] investigated the genetic variations in genes primarily related to oxidative stress and antioxidants in two subject groups, the Lothian Birth Cohort and the Aberdeen Birth Cohort, who were studied at (a) 11 and 79 years and (b) at 11 and 64 years, respectively, when data on lifetime trait and lifetime cognitive changes were recorded. SNPs were identified in 109 genes implicated in oxidative stress and/or cognition. Interestingly, an intronic SNP in the APP gene was significantly associated with cognitive aging when both studies were combined. The nerve growth factor gene showed genetic polymorphism in a Japanese study, the NGF SNP re6330 genotype, and correlated with executive dysfunction.[17]

Epigenetic modifications may act as aging regulators, although whether they play a direct or indirect role remains to be elucidated.

2.3.1 DNA Methylation

Substantial evidence suggests that there is a gradual loss of global DNA methylation during aging (reviewed[18]), possibly *via* a reduction of 5-methylcytosine. This is due to a reduction of DNA methyltransferases (DNMTs). During aging there may be a variable response by these different DNMTs and further studies are needed. Such hypo-methylation would promote genomic instability with the activation of oncogenes. Animal studies have identified a distinct change in the pattern of methylation of the Arc gene (a memory promoting gene) in the dentate gyrus of the aged hippocampus, the aged rats showing less DNA methylation of the Arc gene under basal conditions but increased methylation of the Arc gene following spatial learning.[19]

Caloric restriction may also alter DNA methylation. Studies of prenatal exposure to the Dutch hunger winter at the end of the Second World War was shown to reduce DNA methylation at the imprinted loci of insulin growth factor 2 at least 60 years later[20] and, interestingly, was associated with a higher incidence of coronary heart disease and obesity in later life.[21] Many vertebrate and invertebrate models have been utilized to show that diet can regulate gene expression *via* changes in DNA methylation, *e.g.* the murine agouti viable gene in mice and royal jelly ingestion in honey bees. Caloric restriction, without malnutrition but including essential nutrients, may enhance survival in various animal models. Such caloric restrictions are associated with lower ROS production and less oxidative damage, as well as reduced iron accumulation in rat brain.[22] Whether marginal iron depletion will have an anti-aging paradigm in man remains to be ascertained. The results of various studies which have investigated the association between caloric restriction and longevity, in many different animal models, have been contradictory. Such caloric restrictions may delay the onset of various diseases such as cardiovascular disease, diabetes and cancer[23] and indirectly delay the subsequent onset of cognitive impairment.

Most recently, Mattison *et al.*[24] reported that a caloric restricted diet did not enhance longevity in rhesus monkeys during a 27 year study period.

2.3.2 Histone Post-translational Modifications

Post-translational histone modification has been linked to aging and longevity.[18] Hypermethylation of the promoter of tumour suppressor genes as well as aberrant DNA (cytosine-5-)methyltransferase 1 are linked to aging and cancer. Sirtuins are a family of NAD^+ dependent class III histone deacetylases, and the structure of yeast sirtuin 2 (Sir2) has been determined. Sirtuins regulate important biological pathways in eukaryotes, and can remove the acetyl group from histones as well as other substrates. They have been implicated in aging, since a deletion of Sir2 shortens lifespan while, in contrast, increased gene dosage extends it. It was suggested that the anti-aging effect of Sir2 was mediated *via* its translocation from telomeres and mating-type loci to ribosomal DNA (rDNA). However, more recent studies have indicated a more complex situation. SIRT1, the mammalian ortholog of Sir2, may be linked to longevity *via* its modifying effect on the transcription of NFκB, as well as of the tumour suppressor's p53 and p73. In this way, the expression of inflammatory and stress response genes will be modified, which will indirectly affect cognitive function. The activity of sirtuins are modulated by caloric restriction: less NAD^+ is required for metabolic reactions and the excess NAD^+ will bind to Sir2, enhancing its activity. However, further studies are needed to confirm the relationship between sirtuins and aging.

With aging there is a decrease in transcription of a number of genes which affect the ability of the brain to retain its plasticity. These include immediate early genes (IEGs), *e.g.* Arc, an activity-regulated cytoskeletal gene, Zif268, a nerve growth factor inducible-A and early growth response gene, as well as brain-derived neurotrophic factor (BDNF; reviewed[19]). Animal studies have shown that blocking the expression of these genes will prevent the consolidation of memory. All pathways of the innate immune system are upregulated with aging, which include genes involved in toll-like receptor signalling and inflammasome signalling.[25] Clearly further studies are needed to ascertain whether the genes involved in the establishment of longevity will also influence the maintenance of cognitive function with aging.

2.4 Proteins Possibly Involved in the Disease

As yet, it is not possible to identify subjects with MCI who will progress to Alzheimer's disease or not; the identification of proteins which are implicated solely in MCI, possibly independent of amyloid protein and neurofibrillary tangles, has not been easy to establish. In many studies where MCI is diagnosed, an increase in amyloid proteins is often found, their presence in the brain of asymptomatic elderly patients possibly being associated with accelerated atrophy.[26] Various criteria, such as Monte Carlo selection, have been used to assess risk for future Alzheimer's disease.[27] These include mini-mental state examination scores, apoprotein E genotype and cerebrospinal markers such as total tau, phosph-tau-181 and the 42 amino acid form of amyloid β (Aβ42).

As already eluded to, age-related memory decline is manifested primarily in declarative/episodic and working memory, which is attributable to the hippocampus and prefrontal cortex. In post-mortem brain tissue, hippocampal precursors of nerve growth factor, proNGF/NGF signalling, TrkA and Akt were significantly reduced in MCI compared to non-cognitive impaired subjects.[6] The authors suggested that such alterations in the hippocampal NGF signalling pathway in MCI would favour proNGF mediated pro-apoptotic pathways. Alterations in the relative abundance of synaptic proteins in the hippocampus may be an important factor in MCI. Debrin, a signal transduction molecule, may play a role in hippocampal development and dendritic spine morphogenesis as well as neural mechanisms of cognition. Reduction of postsynaptic debrin, by approximately 40%, occurred in the hippocampus of MCI patients.[28]

2.5 Neuroinflammation

Normal aging is characterized by a chronic low-grade inflammatory state, which is shown by the overexpression of proinflammatory cytokines to the detriment of the anti-inflammatory cytokines and other molecules. This may contribute to the impairment of various aspects of cognitive function, although explanations for such inflammatory changes remain unknown. They are characterized by the chronic activation of perivascular macrophages and microglia which will release proinflammatory cytokines, elevated levels of reactive oxygen and nitrogen species. Astrocyte numbers are also increased. It has been suggested by Dilger and Johnson[29] that there is "priming of microglia" by an unknown process, which makes them more responsive to immune challenges. A challenge with lipopolysaccharide (LPS) in aged mice resulted in overproduction of cytokines, which was associated with cognitive alterations (reviewed[30]). It is therefore possible that chronic activation of microglia in elderly subjects increases their vulnerability to various neuropsychiatric disorders, including MCI. Indeed, activated inflammatory processes, when associated with poor antioxidant status as well as altered tryptophan metabolism, induced reduced quality of life in the elderly.[31] In addition, there was an association between inflammatory status and cognitive performance/decline in overweight and obese women, the latter group showing low-grade inflammatory processes due to the release of proinflammatory cytokines from macrophages accumulating in adipose tissue. (reviewed[30]). People with type 2 diabetes showed elevated circulating levels of proinflammatory cytokines, IL-6 and TNF-α.[32] This was associated with late-life poorer cognitive ability, although it was unclear whether such changes are cofounders or whether they might be mediators of a pathway which leads to cognitive decline. Other studies have also shown that there is a relationship between inflammatory biomarkers and cognition in non-diabetic populations.[33–35] Cognitive defects induced by inflammation are typically observed in the hippocampal-dependent tasks,[36] affecting memory formation, which extends from impaired acquisition of various stages of memory formation to disruption, consolidation and reconsolidation. It might be hoped that this would be a transient effect and, subsequent to cessation of the inflammatory response, an improvement in cognition would result. However, the presence of an underlying inflammation in the elderly would make them more susceptible to persistent inflammation-induced cognitive deficits.

2.6   Metals Involved; Pathways Activated by Metals

It has been established for many years that various metals, notably iron and copper, increase in the aging brain, although the cause of such increased metal accumulation in specific brain regions is undefined. In addition, increases in these redox-active metals may also promote oxidative stress.

2.6.1   Iron and Aging

With aging, brain iron content increases, the iron being contained primarily within the iron storage proteins, H- and L-ferritin, and also in neuromelanin (NM). Such iron will not induce toxicity directly but could increase metabolic stress. Increased amounts of ferritin and NM are deposited in the substantia nigra (SN) with aging,[37] which if in excess, as in PD, may induce oxidative stress.[38] It is of interest that NM is able to bind a number of elements, particularly iron, copper and zinc, such that the presence of this protein could be an important protective factor by binding to a range of toxic ions.[39] NM pigment is not detectable in prenatal or infant neurons, but the size and numbers of these pigment granules increase in certain neurons [with 95% of the dopaminergic neurons of the SN as well as in the noradrenergic neurons of the locus coeruleus (LC) showing deposits] with age.[40] Interestingly, this pigment shows a similar distribution to that of ferritin with aging, remaining stable in the LC but increasing in the SN.[37] Such increases in iron in brain regions with aging may contribute to the low-grade neuroinflammation that is evident in the aging brain, thereby inducing cognitive changes. The iron content of other brain regions in elderly post-mortem brains has also been determined. A correlation between an increase in hippocampal iron and cognitive defects was reported in one study,[41] where a worse verbal-memory performance was associated with higher hippocampal iron in men but not in women. In addition, the worse verbal-memory performance was associated with higher basal ganglia iron in both males and females. Such iron was possibly present in ferritin, since the field-dependent relaxation rate measurement identified ferric oxyhydroxide particles. However, since NM also has a similar chemical structure,[42] this pigment might also be present. In a recent MRI study, when MRI T_2* was utilized to determine iron content, age differences in memory were attributed to smaller hippocampal volumes and higher hippocampal iron concentrations,[43] while no such association was evident in either the caudate nucleus or primary visual cortex. Such studies clearly implicate an association between increases of iron in specific brain regions and cognitive dysfunction.

2.6.2 Copper and Aging

Brain copper homeostasis during aging has not been extensively studied. Various studies have investigated changes in human and mouse brain copper content as well as copper-containing enzymes with age. In one study, the copper content decreased with age in human brains in the LC while no significant changes were evident in the SN. However, the copper content of NM in the LC was higher than the SN, which could indicate higher copper mobilization in LC neurons.[37] Other regions of the brain have not been studied in healthy aged post-mortem material. In contrast, studies in mice have identified regional selective changes in copper brain content with normal aging,[44] with increases in the global cerebral content.[45] However, lower copper contents were noted in the striatum and ventral cortex in aging mice.[45] The focal areas of copper deficit correspond to the regions of greatest reduction in SOD-1 activity in the aged mice. Indeed, there were decreases in SOD-1 activity in brains of male C57BL/6J mice with age by 36% on a protein basis.[44] In human studies, Cu-Zn SOD activity did not vary with aging in either the LC or SN, while Mn SOD decreased only in the LC and not the SN with aging.[46] In a further study, the relationship between levels of Aβ and biological metals in CSF was investigated in Japanese American men ($N = 131$) from the population-based Honolulu–Asia Aging Study. A significant inverse correlation was observed between CSF Aβ42 and copper, zinc, iron, manganese and chromium content.[47] The role played by copper, if any, in the development of mild cognitive impairment is undefined. Some studies have emphasized the importance of free copper present in the blood (reviewed[48]), which could rapidly cross the blood–brain barrier (BBB) and enhance amyloid plaque deposition in the brain, resulting in cognitive decline. An increase in the ratio of serum copper to non-heme iron levels predicted those patients with MCI who would develop AD. However, no changes were evident in MCI patients.[49] A high fat diet combined with a relatively high copper intake was reputed to induce cognitive decline.[50] Clearly more studies are needed to identify the involvement of copper in cognitive decline.

2.6.3 Zinc and Aging

Zinc is essential for the formation and migration of neurons as well as in the functioning of neuronal synapses. Changes in zinc homeostasis in the brain, possibly *via* changes in zinc transporters, could therefore have a detrimental effect on cognition. A marginal increase in the zinc transporter ZnT-6, a protein responsible for sequestration of zinc in the trans-Golgi network, in the hippocampus/parahippocampal gyrus may play a role in amnestic mild cognitive impairment.[51]

2.7 Neurotransmitters Implicated in MCI

During normal aging, there is a decrease in the excitability of cortical and hippocampal neurons which may cause altered network functioning of neural circuits in brain regions that are critical for storing and retrieving memory. In addition, neurogenesis, which occurs in the hippocampus throughout life, may show a decline in the production of the adult progenitor and neural cells with aging, which could contribute to the cognitive deficits. The excitatory neurotransmitter glutamate and its receptor play an important role in spatial learning and other hippocampal-dependant memory processes. Ionotrophic glutamate receptors play a critical role in synaptic plasticity, while group 1 metabotrophic glutamate receptors are also involved in cognitive aging, the latter being enriched in the hippocampal formation as well as interacting with other proteins in the membrane, including ionotropic receptors (Figure 2.6). Synaptic plasticity is important for the maintenance of cognition, as well as long-term depression which is induced by group 1 metabotrophic glutamate receptors. Simulation of the ionotrophic glutamate receptors, including AMPA and NMDA receptors, induces an influx of Ca^{2+} into the cell, which will activate a variety of intracellular signalling pathways to induce synaptic plasticity and long-term potentiation, *i.e.* long-lasting signal enhancement (reviewed[52]). Changes in calcium homeostasis have been associated with age-related memory impairment.

Cholinergic neurons within the nucleus basalis and the septal diagonal band complex provide the major source of cholinergic innervation to the cerebral cortex and hippocampus, respectively, and play an important role in memory and attention function. Two types of receptors—muscarinic and nicotinic—

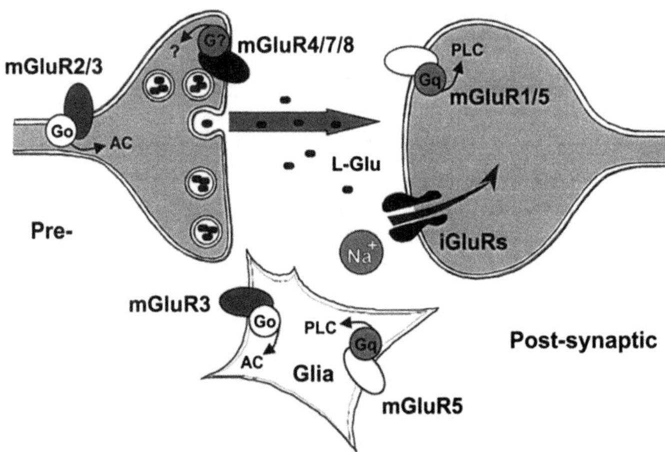

Figure 2.6 Involvement of glutamate receptors in cognitive aging. L-Glutamate acts as the chemical transmitter for excitatory signals. It is released into the synaptic cleft and activates a multitude of highly integrated pathways by binding to iono- and metabotrophic glutamate receptors.

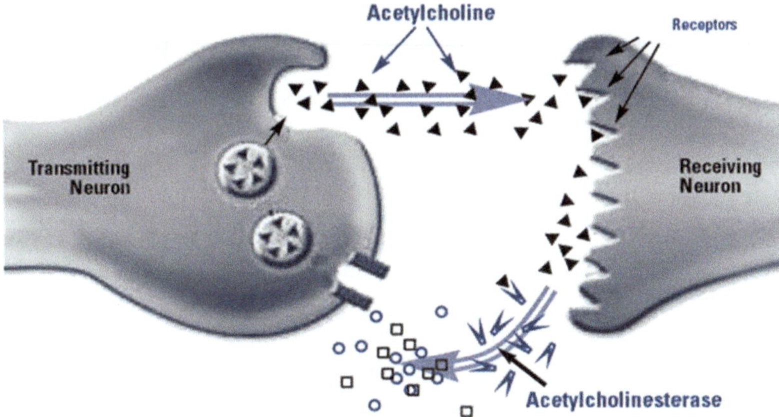

Figure 2.7 Acetylcholine neurotransmission. After signalling, acetylcholine is released from its receptors, broken down by acetylcholinesterase and recycled in a continuous process. Restoration of cholinergic function may reduce the severity of the cognitive loss. This hypothesis has been supported by the finding that cholinesterase inhibitors show positive effects on cognition. Acetylcholinesterase inhibitors will suppress acetylcholinesterase activity, thereby preventing it from degrading acetylcholine, and enhancing acetylcholine levels within the synapse. Two acetylcholinesterase inhibiting drugs currently approved for AD (Tacrine™ and Aricept™) are only moderately effective, show side effects such as liver toxicity and are expensive. However, in normal aging, such cholinesterase inhibitors have not been successful in reversing normal age-related cognitive deficits.
(Reproduced with permission from www.stoolaf.edu).

respond to acetylcholine to facilitate intracellular communication, memory processing and higher cognitive functions; acetylcholine is released into the synapse (Figure 2.7), where it binds to a receptor on the post-synaptic terminal. Once the signal is triggered, acetylcholine is rapidly broken down by acetylcholinesterase and the breakdown products made available for recycling. Diminished cholinergic functioning, a biomarker of normal aging, is especially severe in cases involving dementia. Degeneration of the cholinergic basal forebrain neurons correlate with dementia severity, disease duration and cognitive impairment (Figure 2.8). The viability of the cholinergic basal forebrain neurons is dependent on nerve growth factor which is transported *via* a complex interaction between two receptors, the high affinity nerve growth factor (NGF) receptor, also known as TRK1-transforming tyrosine kinase protein, TrkA, and the putative cell death associated low-affinity pan neuroptrophin receptor.[7] NGF receptor TrkA, which is expressed in cholinergic forebrain neuronal populations of basal forebrain and striatum, is markedly reduced in individuals with MCI without dementia and early-stage AD. Sanchez-Otis *et al.*[53] identified a key role for TrkA signalling in establishing the basal forebrain cholinergic circuitry through the extracellular signal-regulated kinase (ERK) pathway. The normal developmental increase of

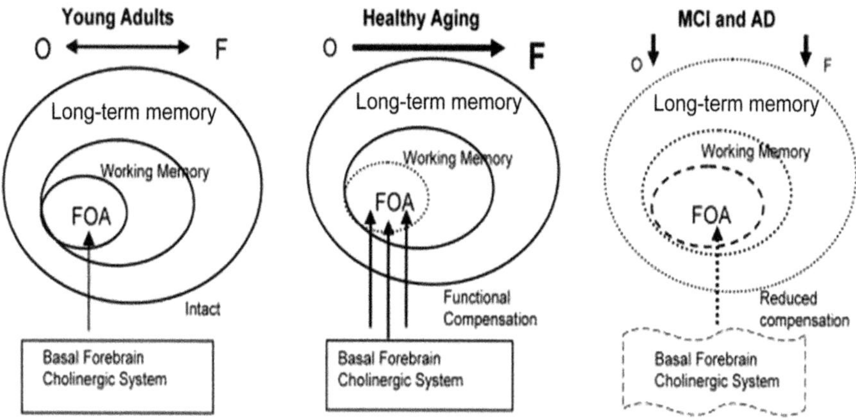

Figure 2.8 Cholinergic functional compensation model of age-related cognitive dysfunction. The figure shows the functional compensation model illustrating the effects of aging and cholinergic dysfunction. *Dotted* and *grey lines* indicate impairments. When younger adults (*left panel*) perform attention and memory tasks, they show brain activation patterns that are balanced between occipital and frontal regions, thus allowing for a balance of bottom-up and top-down processes. Normal cognitive aging (*middle panel*) may degrade the control processes of the focus of attention, thereby affecting working memory and long-term memory. Functional compensation will recruit the cholinergic system and associated cortical areas to maintain good performance on a task. We propose that this cholinergic recruitment will result in increases in frontal activation for older adults (PASA). Cholinergic dysfunction seen in MCI and AD (*right panel*) will lead to further attentional control impairments that cannot be compensated for by increased recruitment of the cholinergic system, causing working memory and long-term memory processes to be impaired further. The activation patterns will show decreases in frontal and occipital regions. Underlying mechanisms through which physical activity may exert neuro-protective effects: $\uparrow$, increased; $\downarrow$, decreased; IGF, insulin-like growth factor; BDNF, brain-derived neurotrophic growth factor; VEGF, vascular endothelial growth factor. Is there a role for physical activity in preventing cognitive decline in people with mild cognitive impairment? (Reproduced from Dumas and Newhouse[83] with permission from Elsevier).

choline acetyltransferase (ChAT) expression is critically dependent on TrkA signalling before neuronal connections are established. Such a lack of TrkA signalling will have a selective impact on cognitive activity.

Other neurotransmitter systems are involved in cognitive dysfunction, *e.g.* cognitive impairment has been associated with down-regulation in the biosynthesis of *N*-acetylserotonin in rabbits, which is caused by up-regulation of β-adrenoceptors and the availability of serotonin.[54] Clearly, within the next few years the role played by a number of other neurotransmitter systems, *e.g.* the GABA and dopamine systems, will be elucidated.

2.8   Therapeutics

Currently there are no approved pharmacological treatments for MCI. There may be a reduced risk of MCI in subjects taking antihypertensive medications, cholesterol-lowering drugs, antioxidants, anti-inflammatories and oestrogen therapy, although there have been no placebo controlled clinical trials of these associations. Currently, the only advice to the aging population is to maintain a healthy lifestyle, with exercise and good nutrition. Strategies to reduce overweight and obesity in the population may reduce low-grade proinflammatory processes which seem to be involved in the development of MCI. High homocysteine, APOE ε4 allele, heart disease, poor odour identification and lower mental activity are biomarkers for MCI.[55] Asymptomatic atherosclerotic disease and cardiovascular risk factors predict a decline in cognitive function.

2.8.1   Exercise

Physical exercise has many beneficial effects on brain function: increasing the activity of several neurotrophic and vascular growth factors, including insulin-like growth factor-1, brain-derived neurotrophic factor and vascular endothelial growth factor (Figure 2.9; (reviewed[56]), neurogenesis, angiogenesis, synaptic plasticity and dendritic spine density in the hippocampus. It is clear that such increased physical exercise will also have additional positive influences on various

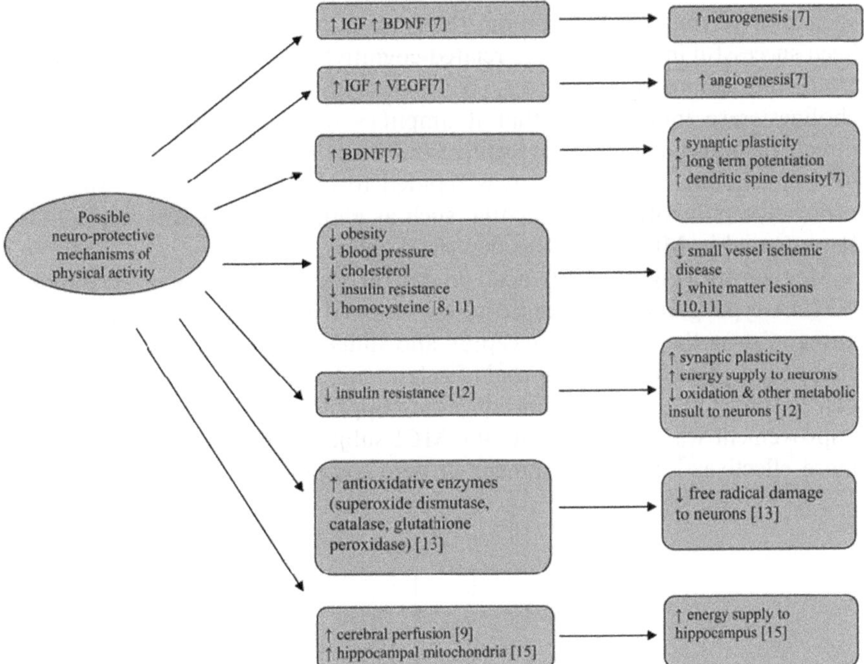

Figure 2.9   Beneficial aspects of exercise to prevent MCI.

cardiovascular risk factors such as blood pressure, cerebral blood flow, obesity and hypercholesterolaemia, as well as improving insulin resistance and increasing synaptic plasticity and energy metabolism. Physical exercise may also enhance the expression of genes that regulate the production of free-radical scavenging enzymes, as well as increasing production of mitochondria in neurons, thereby enhancing energy metabolism in the brain (reviewed[56]). In one study of cognitively normal older people it was shown that there was greater grey and white matter volume (measured by MRI) in people with higher aerobic capacity compared with those with lower aerobic capacity (reviewed[56]). Cybercycling (stationary cycling with virtual reality tours) by older adults showed better cognitive function than traditional exercise, for the same effort, indicating that simultaneous cognitive and physical exercise has greater potential for preventing cognitive decline. An enhanced neuroplasticity was also observed in this study, caused by raised plasma brain-derived neurotrophic factor.[57]

2.8.2 Neurotransmitter System Modifiers

Restoration of cholinergic function may reduce the severity of the cognitive loss. This hypothesis has been supported by the finding that cholinesterase inhibitors show positive effects on cognition. Acetylcholinesterase inhibitors will suppress acetylcholinesterase activity, thereby preventing acetylcholine degradation and enhancing acetylcholine levels within the synapse. However, two acetylcholinesterase inhibiting drugs currently approved for the treatment of AD (Tacrine™ and Aricept™) show limited efficacy and induce side effects, such as liver toxicity. In addition, their use in normal aging subjects has not been successful in reversing age-related cognitive deficits. Galantamine derived from the common snowdrop (*Galanthus nivalis*) is an effective acetyl-cholinesterase inhibitor, in that it amplifies the effects of acetylcholine by directly stimulating nicotinic receptors (nAChR), thereby increasing the release of acetylcholine. In addition, it is reputed to modulate levels of other neur-otransmitters involved in dementia, such as glutamate, serotonin and GABA. Its potential in MCI subjects is described below.

Memantine (Figure 2.10) acts on the glutamatergic system by blocking NMDA-type glutamate receptors, as well as acting as a non-competitive antagonist at the serotonin receptor and different neuronal nicotinic acetyl-choline receptors. In a clinical trial of galantamine +/− memantine in a two-year placebo controlled study, safety concerns stopped the trial although a significant improvement was noted in amnestic MCI subjects, the combined drugs being more effective.[58] The development of new compounds to modify changes in

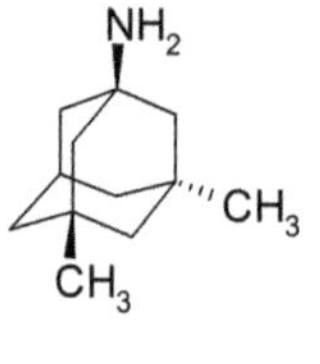

Figure 2.10 Chemical structure of memantine.

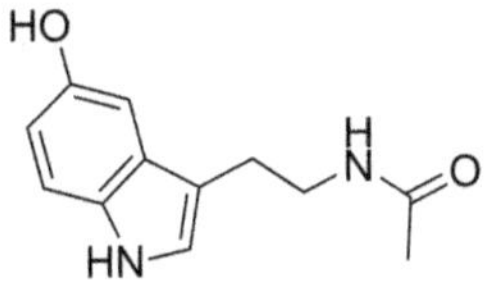

Figure 2.11   Chemical structure of *N*-acetylserotonin.

neurotransmitter concentration and their neurochemical action is needed. Down-regulation of the biosynthesis of *N*-acetylserotonin (NAS), a methoxy-indole derivative of tryptophan, may be involved in age-associated cognitive impairment (Figure 2.11). Aging is indeed associated with decreased NAS production, and NAS is reputed to exert antidepressant-like and cognition-enhancing effects. NAS (and its derivatives) will attenuate attenuated cognitive impairment induced by cholinergic neurotoxins, as well as protecting against β-amyloid neurotoxicity (reviewed[59]). Since NAS is a potent agonist of the high-affinity BDNF tyrosine kinase (TrkB) receptors, the antidepressant and cognition-enhancing effect of NAS might be mediated by activation of TrkB receptors. NAS may also have anti-inflammatory and anti-oxidative effects. In animal studies, three weeks administration of NAS reversed the impairment of performance in active avoidance and water maze tests induced by cholinergic neurotoxins,[60] while its administration subcutaneously decreased oxidative stress in the brain of 11-month-old mice. This might be due to stimulation of glutathione peroxidase, suppression of phospholipase A2 or inhibition of sepiapterin reductase, the key enzyme for the biosynthesis of tetrahydrobiopterin, the essential co-factor of nitric oxide synthase.[61] The use of beta-blockers and benzodiazepines will negatively affect cognitive function and NAS synthesis.

2.8.3   Smoking

Longitudinal studies, which have investigated an association between various vascular risk factors with cognitive function, have produced contradictory results (reviewed[62]). A recent study by Dregan *et al.*,[62] when over 8000 individuals over 66 years were investigated over an eight-year period, a greater risk of cognitive decline, with respect to global cognition, memory and executive scores, was identified in smokers compared with non-smokers at both follow-up stages of four and eight years. In addition, the longitudinal study also identified lower global cognitive and specific memory in subjects with blood systolic pressure >160 mmHg. It is clear that such cardiovascular risk factors, as well as smoking, are modifiable risk factors for cognitive decline.

2.8.4   Diet

Diet is an important factor in the aging process; epidemiological studies have identified certain foods which may be able to diminish the decline in cognition. Such supplements, often termed nutriceuticals, may provide protection as well as diminishing the progression of mild cognitive impairment.

2.8.4.1 *Saturated/unsaturated Fat Intake*

The consumption of specific types of fat, rather than total fat intake itself, appears to influence cognitive aging. In one study, 39 876 female health professionals were recruited into a study between 1992 and 1995, with cognitive testing beginning at the age of 65 years. Over the next four years, 89% completed the initial assessment, while 85% undertook further assessment in 2000 and 82% in 2002. The results showed that community-dwelling older women, who consumed a diet high in saturated fats, were associated with a poorer cognition and verbal memory over the series of assessments. In contrast, a higher monounsaturated fat intake was related to better global cognitive and verbal memory recall.[63] In another study the effect of high calorific intake on the occurrence of MCI was studied.[64] The risk of developing MCI was elevated in subjects with high carbohydrate intake but was reduced in subjects with high fat intake and high protein intake. The study concluded that a diet high in calories from carbohydrates and low in fat and protein content could enhance MCI risk. Diet, as well as mild stress, may play an important part in whether the immune response is able to react accordingly with the development of aging.

2.8.4.2 *Polyunsaturated Fatty Acids*

Long-chain omega-3 polyunsaturated fatty acids are involved in many cellular processes in the brain, and their deficiency may affect cognition. Structurally, PUFA are key components of phospholipids, comprising cellular and intra-cellular membranes. Long-chain PUFA are required for the synthesis of eicosanoids, which are important signaling hormones with numerous complex functions. Those derived from omega-3 long-chain PUFA are generally anti-inflammatory, anti-thrombotic and vasodilatory, and may prevent vascular dementia *via* action on lipids, inflammation, thrombosis and vascular function. Docosahexaenoic acid, together with other long-chain omega-3 PUFA (*e.g.* eicosapentaentoic acid), are present in the retina, brain and nervous system as well the neural synapses. With aging, neural membrane fluidity may be compromised, caused by an increase in cholesterol, reduced activities of desaturase enzymes and impaired phospholipid metabolism, as well as oxidative stress. In such circumstances, supplementation with PUFA may be of therapeutic benefit. Epidemiologic studies have generally supported a protective association between fish and omega-3 FA levels and cognitive decline.[65] Some of the small, short-term, randomized trials of docosahexaenoic acid and/or eicosapentaentoic acid supplementation have found positive effects on some aspects of cognition in older adults who were cognitively intact or had mild cognitive impairment.

2.8.4.3 *Berries*

Since stress sand inflammation are major contributors for cognitive decline, the consumption of foods with high content of flavonoids, which act both as powerful antioxidants and anti-inflammatory substances, could be beneficial. Berries are high in flavonoids, particularly anthocyanidins, and have been

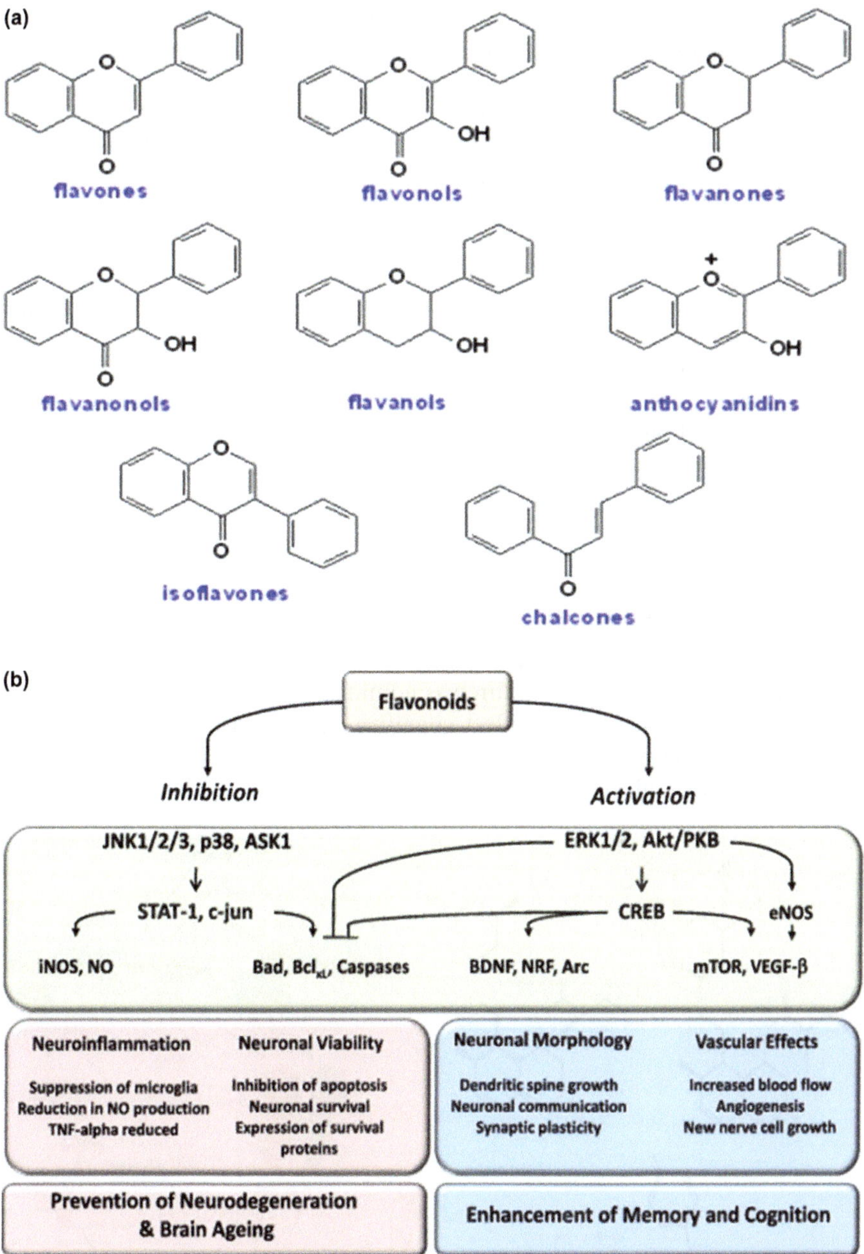

Figure 2.12   (a) Chemical structure of flavonoids. (b) Beneficial effects of flavonoids.

shown to improve cognition in experimental studies (Figure 2.12a). In a study by Devore *et al.*[66] the association between long-term intake of berries and flavonoids and cognition was evaluated in a large, prospective cohort of older women in the Nurses' Health Study. In 1976, 121 700 registered female nurses aged 30–55 years were recruited for the study and responded to mailed questionnaires on their health and lifestyle. Follow-up questionnaires were sent biennially, with a food frequency questionnaire which was updated every four years. Between 1995 and 2001, women aged 70 years and older participated in a telephone-based study of cognitive function over the next four years. Blueberries and strawberries were the primary foods contributing to anthocyanidin intake in this study, while apples and oranges were the main contributors to other flavonoid subclasses and total flavonoid intake.[66] The results showed that subjects who regularly ate high amounts of strawberries and blueberries could diminish the mental decline of old age, by up to two and half years. In addition, greater intakes of anthocyanidins and total flavonoids were also associated with less cognitive deterioration. Consumption of flavonoid-rich foods, green tea or blueberries may benefit cognition by their specific interaction with cellular and molecular targets, *e.g.* ERK and P13-kinase/Akt signalling pathways, increasing blood flow and the ability to initiate neurogenesis in the hippocampus (Figure 2.12b).[67]

2.8.4.4 Statins

Since both coronary and cerebrovascular events are linked to cognitive dysfunction, the administration of statins to prevent their occurrence could be of therapeutic efficacy. Pravastatin was administered to subjects aged between 70 and 82 years, but did not affect cognitive decline during the study period (Figure 2.13).[68] However, if statins were administered at a lower age, this may have a protective effect later in life.

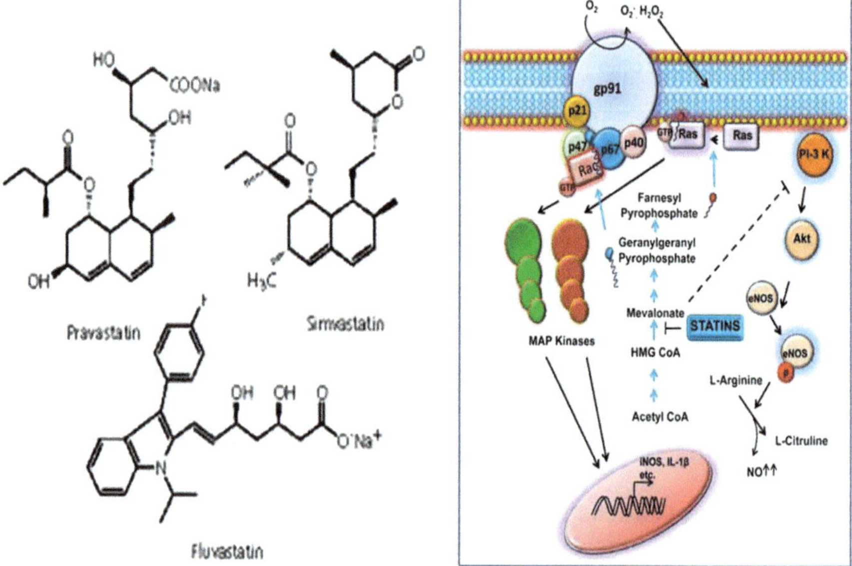

Figure 2.13 Chemical structure of statins and their biological action.

2.8.4.5   Creatine Supplementation

Creatine, a nitrogenous organic acid (Figure 2.14), helps to store energy to all cells in the body, in the form of phosphocreatine, thereby elevating the ability of brain cells to resynthesize ATP from ADP when required. The creatine/ phosphocreatine system is regulated by the mitochondrial creatine kinase, which plays an important role in maintaining energy balance in the brain. In early animal experiments, creatine supplementation improved health and survival in aged mice.[69] Several groups have shown that short-term high-dose creatine supplementation improved cognitive ability in the elderly.[70]

2.8.4.6   Sirtuins

The SIRT1 protein may play a potential role in longevity, as it senses and communicates the energy status of a cell to key mechanisms of mitochondrial regulation and energy production. These mechanisms include the biogenesis of mitochondria, the clearance of damaged organelles and the physiological rhythmicity of gene expression. Underexpression of a human sirtuin gene seems to be detrimental for longevity. NAD-dependent deacetylase sirtuin-3, SIRT3, is located in the mitochondrial matrix. Overexpression of SIRT3 in cultured cells increases respiration and decreases the production of reactive oxygen species. A strong association between SIRT3 alleles and longevity in males has been reported. It is suggested that both SIRT1 and SIRT3 may play a role in regulating the acetylation status and other cellular activities, thereby acting as potential regulators of aging. Resveratrol, a polyphenol molecule present in grapes, activates the SIRT1 gene and extends the lifespan of various species, from yeast to rodents.

2.8.4.7   Bacopa monnieri

Bacopa monnieri, an Indian traditional medicinal plant (Figure 2.15), reversed memory impairment in cochicine-treated rats, by attenuating oxidative damage as well as the activity of the membrane-bound enzymes Na^+-K^+-ATPase and AchE.[71]

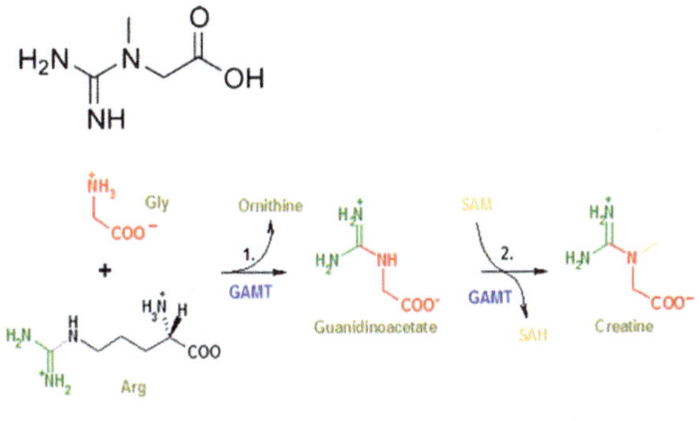

Figure 2.14   Creatine, a nitrogenous organic acid.

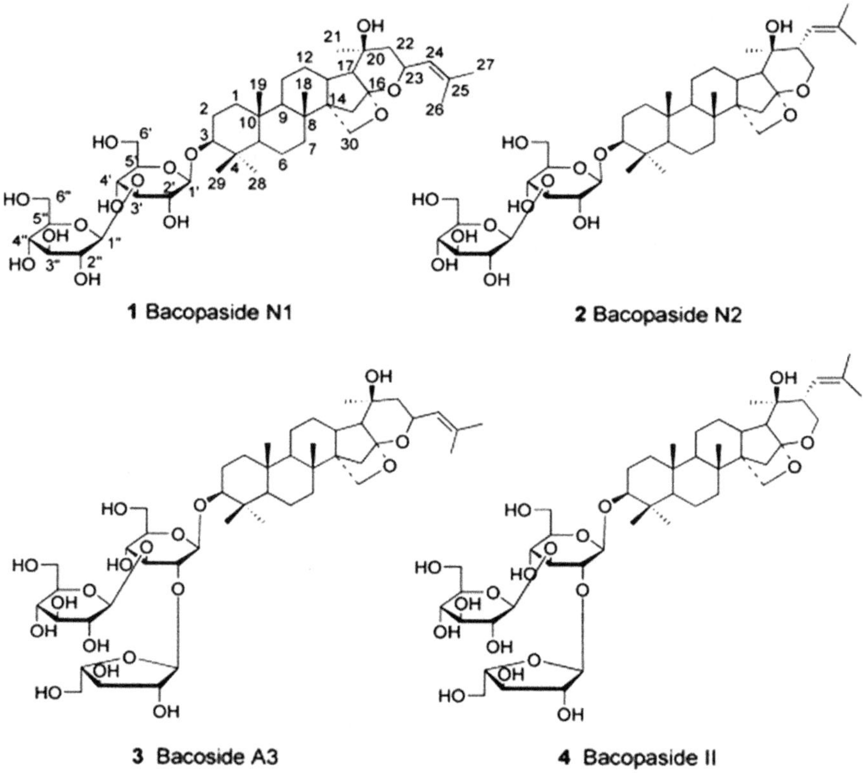

1 Bacopaside N1

2 Bacopaside N2

3 Bacoside A3

4 Bacopaside II

Figure 2.15   Chemical constituents of *Bacopa monnieri*.

2.8.5   Immunity

Immunological age is an important factor: the aging of the immune system will affect the ability of the individual to respond to a given stimulus, as well as to cognitive function and decline. It is suggested that some people age better, because of their ability to maintain their immune system. This will be in part reflected by genetic factors as well as nutrition and exercise through their life span. Age-related memory decline is characterized by marked demyelination and loss of oligodentrocytes in white matter, which could possibly be alleviated by various activities such as physical, social and sensory stimuli.[72] Participation in cognitively stimulating leisure activity will attenuate the effect of white matter lesion pathology.[73]

Targeting the increased levels of cytokines presents an attractive hypothesis, and drugs which will inhibit TNFα and IL-1 or COX1 and COX2, respectively, could be of interest (reviewed[74]). In addition, Capuron[74] extolled the benefits of vitamin E supplementation. In a study of 69 elderly subjects, an association between vitamin E status, immune processes and quality of life in the elderly was identified. Indeed, elderly patients with high plasma concentrations of vitamin E exhibited lower plasma levels of inflammation as well as showing better health status, with respect to both mental and physical quality of life questionnaires.

Exercise will enhance the immune function (specifically T-cell immunity), which will facilitate cognitive ability and improve mental stability, all being mediated by elevation in growth factor production and neuronal plasticity. Boosting T-cell immunity may be beneficial for the associated memory impairment.[75] The T-cells, on encountering their specific antigen presented by the antigen presenting cells, can produce protective compounds such as growth factors.

2.8.6 Mitochondrial Mutations

Oxidative mitochondrial decay occurs with aging. This can be reversed to a limited extent by the administration of large doses of acetylcarnitine or lipoic acid. Indeed, in a recent clinical trial of subjects with MCI or mild AD, acetylcarnitine supplementation showed a beneficial effect, particularly in respect of slowing the progression of the disease. Its mode of action is possibly *via* the enhancement of phospholipid precursors for membrane synthesis as well as reducing mtDNA deletions (reviewed[76]). It is well established that mtDNA accumulates mutations with aging, particularly large-scale deletions and point mutations which may be associated with a decline in mitochondrial function. It is important that the extensive network of antioxidant defences, *e.g.* catalase, glutathione peroxidase and superoxide dismutase, are able to control the release of excessive ROS from the electron transport chain. Therefore enhancing mitochondrial oxidative defences may increase longevity. Profiling of mRNA from post-mortem frontal cortex of subjects aged between 26 and 60 years identified a decrease in expression of genes involved in synaptic plasticity, vesicular transport and mitochondrial function, while there was an increase in the expression of stress responses, antioxidant and DNA-repair genes (reviewed[77]). These results indicate that mitochondrial dysfunction plays an important role in the aging brain.

2.9 Tests for MCI

As yet, there are no tests to demonstrate conclusively the existence of MCI. Various tests, *e.g.* Petersen and CDR tests, which assess memory, planning, judgement, ability to understand visual information and other key thinking skills, are available and may aid diagnosis. Biochemical markers, such as platelet cytochrome C oxidase and glutamine synthetase-like protein, may be early markers of cognitive decline,[78] while leukocyte telomere length was reduced in MCI subjects who did not progress to dementia.[79] Increased levels of plasma IL-6, in combination with other biomarkers, identified subjects with early amnesia MCI,[80] while raised circulating levels of CRP and fibrinogen, as well as elevated plasma viscosity, predicted poorer cognitive ability in old age.[81]

For too long, the occurrence of cognitive decline in aging subjects has been an acceptable result of longevity. As we learn more about the various processes that contribute to this dysfunction, it will be hoped that, within 10 years, supplements which will target specific metabolic processes within specific brain regions will be available, such that longevity will be associated with normal cognitive function.

References

1. I. Reinvang, R. Grambaite and T. Espeseth, *Int. J. Alzheimer's Dis.*, 2012; doi: 10.1155/2012/936272.
2. J. Spaniol, D. J. Madden and A. Voss, *J. Exp. Psychol.: Learn. Mem. Cogn.*, 2006, **32**, 101.
3. H. Eichenbaum, A. P. Yonelinas and C. Ranganath, *Annu. Rev. Neurosci.*, 2007, **30**, 123.
4. R. C. Petersen, J. E. Parisi, D. W. Dickson, K. A. Johnson and D. S. Knopman, *et al.*, *Arch. Neurol.*, 2006, **63**, 665.
5. W. R. Markesbery, *J Alzheimer's Dis.*, 2010, **19**, 221.
6. E. J. Mufson, B. He, M. Nadeem, S. E. Perez and S. E. Counts, *et al.*, *J. Neuropathol. Exp. Neurol.*, 2012, **71**, 1018.
7. L. Vana, N. M. Kanaan, I. C. Ugwu, J. Wuu, E. J. Mufson and L. I. Binder, *Am. J. Pathol.*, 2011, **179**, 2533.
8. N. Raz, in *The Handbook of Aging and Cognition*, F. I. Craik and T. A. Salthouse, Erlbaum, Mahwah, NJ, 2000, pp. 1–90.
9. N. Raz, U. Lindenberger, K. M. Rodrigue, K. M. Kennedy, D. Head and A. Williamson, *et al.*, *Cereb. Cortex*, 2005, **15**, 1676.
10. S. W. Davis, N. A. Dennis, S. M. Daselaar, M. S. Fleck and R. Cabeza, *Cereb. Cortex*, 2008, **18**, 1201.
11. E. V. Sullivan and A. Pfefferbaum, *Neurosci. Biobehav. Rev.*, 2006, **30**, 749.
12. S. Walter, G. Atzmon, E. W. Demerath, M. E. Garcia and R. C. Kaplan, *et al.*, *Neurobiol. Aging*, 2011, **32**, 2109.
13. J. Deelen, M. Beekman, H. W. Uh, Q. Helmer and M. Kuningas, *et al.*, *Aging Cell*, 2011, **10**, 686.
14. C. K. Lee, R. Weindruch and T. A. Prolla, *Nat. Genet.*, 2000, **25**, 294.
15. C. H. Jiang, J. Z. Tsien, P. G. Schultz and Y. Hu, *Proc. Natl. Acad. Sci. U. S. A.*, 2001, **98**, 1930.
16. S. E. Harris, H. Fox, A. F. Wright, C. Hayward and J. M. Starr, *et al.*, *BMC Genet.*, 2007, **2**, 43.
17. T. Nagata, S. Shinagawa, K. Nukariya, R. Nakayama, K. Nakayama and H. Yamada, *Dement. Geriatr. Cogn. Disord.*, 2011, **32**, 379.
18. C. Huidobro, A. F. Fernandez and M. F. Fraga, *Mol. Aspects Med.*, 2012, 22771540.
19. K. Kosik, P. R. Rapp, N. Raz, S. A. Small, J. D. Sweatt and L.-H. Tsai, *J. Gerontol.*, 2012, **67**, 741.
20. B. T. Heijmans, E. W. Tobi, A. D. Stein, H. Putter and G. J. Blauw, *et al.*, *Proc. Natl. Acad. Sci. U. S. A.*, 2008, **105**, 17046.
21. R. C. Painter, T. J. Roseboom and O. P. Bleker, *Reprod Toxicol*, 2005, **20**, 345.

22. C. I. Cook and B. P. Yu, *Mech. Ageing Dev.*, 1998, **102**, 1.
23. R. J. Colman, R. M. Anderson, S. C. Johnson, E. K. Kastman and K. J. Kosmatka, *et al.*, *Science*, 2009, **325**, 201.
24. J. A. Mattison, G. S. Roth, T. M. Beaseley, E. M. Tilmont and A. M. Handy, *et al.*, *Nature*, 2012, **489**, 318.
25. D. H. Cribbs, N. C. Berchtold, V. Perreau and P. D. Colemen, *et al.*, *J. Neuroinflammation*, 2012, **9**, 179.
26. G. Chetelat, V. L. Villemagne, N. Villain, G. Jones, K. A. Ellis and D. Ames, *et al.*, *Neurology*, 2012, **78**, 477.
27. M. Kruczyk, H. Zetterberg, O. Hansson and S. Roistad, *et al.*, *J. Neural. Transm.*, 2012, **119**, 821.
28. S. E. Counts, B. He, M. Nadeem and J. Wuu, *et al.*, *Neurodeger. Dis.*, 2012, **10**, 216.
29. R. N. Dilger and R. W. Johnson, *J. Leukoc. Biol.*, 2008, **84**, 932.
30. L. Capuron and A. H. Miller, *Pharmacol. Ther.*, 2011, **130**, 226.
31. L. Capuron, A. Moranis, N. Combe, F. Cousson-Gelie, D. Fuchs, V. De Smedt-Peyrusse, P. Barberger-Gateau and S. Laye, *Br. J. Nutr.*, 2009, **102**, 1390.
32. R. E. Marioni, M. W. J. Strachan, R. M. Reynolds, G. D. O. Lowe and R. J. Mitchell, *et al.*, *Diabetes*, 2010, **59**, 710.
33. S. B. Rafnsson, I. J. Deary, F. B. Smith, M. C. Whiteman, A. Rumley and G. D. Lowe, *J. Am. Geriatr. Soc.*, 2007, **55**, 700.
34. M. Luciana, R. E. Marioni, A. J. Gow, J. M. Starr and I. J. Deart, *Psychosom. Med.*, 2009, **71**, 404.
35. P. C. Elwood, J. Pickering and J. E. Gallacher, *Age Ageing*, 2001, **30**, 135.
36. R. A. Kohman and J. S. Rhodes, *Brain Behav. Immun.*, 2013, **27**, 22.
37. F. A. Zucca, C. Bellei, S. Giannelli, M. R. Terreni and M. Gallorini, *et al.*, *J. Neural. Transm.*, 2006, **113**, 757.
38. D. T. Dexter, A. E. Holley, W. D. Flitter, T. F. Slater, F. R. Wells, S. E. Daniel and A. J. Lees, *et al.*, *Ann. Neurol.*, 1992, S82.
39. R. R. Crichton and R. J. Ward, *Curr. Med. Chem.*, 2003, **10**, 997.
40. K. L. Double, V. N. Dedov, H. Fedorow, E. Kettle and G. M. Halliday, *et al.*, *Cell. Mol. Life Sci.*, 2008, **65**, 1669.
41. G. Bartzokis, P. H. Lu., K. Tingus and D. G. Peters, *et al.*, *Neuropsychopharmacology*, 2011, **36**, 1375.
42. M. Gerlach, A. X. Trautwein, L. Zecca, M. B. Youdim and P. Riederer, *J. Neurochem.*, 1995, **65**, 923.
43. K. M. Rodigue, A. M. Daugherty, E. M. Haacke and N. Raz, *Cereb. Cortex*, 2012, 22645251.
44. H. R. Massie, V. R. Aiello and A. A. Iodice, *Mech. Ageing. Dev.*, 1979, **10**, 93.
45. L. M. Wang, J. S. Becker, Q. Wu, M. F. Oliveira and F. A. Bozza, *et al.*, *Metallomics*, 2010, **2**, 348.
46. L. Zecca, M. Gallorini, V. Schunemann and A. X. Trautwein *et al.*, *J Neurochem.* 2001, **76**, 1766.
47. D. Strozyk, L. J. Launer, P. A. Adlard and R. A. Cherny, *et al.*, *Neurobiol. Aging*, 2009, **30**, 1069.

48. G. J. Brewer, *Am. J. Nutr.*, 2009, **28**, 238.
49. C. Mueller, M. Schrag, A. Crofton, J. Stolte, M. U. Muckenthaler, S. Magaki and W. Kirsch, *J. Alzheimer's Dis.*, 2012, **29**, 341.
50. M. C. Morris, D. A. Evans, C. C. Tangney, J. L. Benias and J. A. Schneider, *et al.*, *Arch. Neurol.*, 2006, **63**, 1085.
51. M. A. Lovell, J. L. Smith and W. R. Markesbery, *J. Neuropathol. Exp. Neurol.*, 2006, **65**, 489.
52. C. Menasrd and R. Quirion, *Front. Pharmacol.*, 2012, **3**, 182.
53. E. Sanchez-Ortiz, D. Yui, D. Song, Y. Li and J. L. Rubenstein, *et al.*, *J. Neurosci.*, 2012, **32**, 4065.
54. G. Oxenkrug and R. Ratner, *Aging Dis.*, 2012, **3**, 330.
55. P. S. Sachdev, D. M. Lipinicki, J. Crawford, S. Reppermund and N. A. Kochan, *et al.*, *J. Am. Geriatr. Soc.*, 2012, **60**, 24.
56. S. E. Barber, A. P. Clegg and J. B. Young, *Age Ageing*, 2012, **41**, 5.
57. C. Anderson-Hanley, P. J. Arciero, A. M. Brickman and J. P. Nimon, *et al.*, *Am. J. Prev. Med.*, 2012, **42**, 109.
58. O. Peters, D. Lorenz, A. Fesche, K. Schmidtke, M. Hull, R. Perneczky and E. Ruther, *et al.*, *J. Nutr. Health Aging*, 2012, **16**, 544.
59. G. Oxenkrug and R. Ratner, *Aging Dis.*, 2012, **3**, 330.
60. S. Bachurin, G. Oxenkrug and N. Lermontova, *et al.*, *Ann. N. Y. Acad. Sci.*, 1999, **890**, 155.
61. G. Oxenkrug, *Ann. N. Y. Acad. Sci.*, 2005, **1053**, 334.
62. A. Dregan, R. Stewart and M. C. Gulliford, *Age Ageing*, 2013, **42**, 338.
63. O. I. Okereke, B. A. Rosner, D. H. Kim, J. H. Kang, N. R. Cook and J. E. Manson, *et al.*, *Ann. Neurol.*, 2012, **72**, 124.
64. R. O. Roberts, L. A. Roberts and Y. E. Geda, *J. Alzheimer's Dis.*, 2012, **32**, 329.
65. N. Sinn, C. Milte and P. R. Howe, *Nutrients*, 2010, **2**, 128.
66. E. E. Devore, J. H. Kang, M. M. Breteler and F. Grodstein, *Ann. Neurol.*, 2012, **72**, 135.
67. R. J. Williams and J. P. E. Spencer, *Free Radical Biol. Med.*, 2012, **52**, 35.
68. S. P. Mooijaart, N. Sattar, S. Trompet and E. Polisecki, *et al.*, *PLoS One*, 2011, **6**, e23890.
69. A. Bender, J. Beckers, I. Schneider, S. M. Holter, T. Haack and T. Ruthsatz, *et al.*, *Neurobiol. Aging*, 2008, **29**, 1404.
70. E. S. Rawson and A. C. Venezia, *Amino Acids*, 2011, **40**, 1349.
71. N. Saini, D. Singh and R. Sandhir, *Neurochem. Res.*, 2012, **37**, 1928.
72. S. Yang, W. Lu, D. S. Zhou and Y. Tang, *Anat. Rec. (Hoboken)*, 2012, **259**, 1406.
73. J. S. Saczynski, M. K. Jonsdottir, S. Sigurdsson and G. Eiriksdottir, *et al.*, *J. Gerontol., Ser. A*, 2008, **63**, 848.
74. L. Capuron and A. H. Miller, *Pharmacol. Ther.*, 2011, **130**, 226.
75. N. Ron-Harel and M. Schwartz, *Trends Neurobiol.*, 2009, **32**, 367.
76. B. N. Ames and J. Liu, *Ann. N. Y. Acad. Sci.*, 2004, **1033**, 108.
77. M. T. Lin and M. F. Beal, *Nature*, 2006, **443**, 787.

78. G. Sh. Burbaeva, I. S. Boksha, O. K. Savushkina and M. S. Turishcheva, *et al.*, *Zh. Nevrol. Pskikhiatr. Im S. S. Korsakova*, 2012, **112**, 55.
79. S. Moverare-Skrtic, P. Johansson, N. Mattsson, O. Hansson and A. Wallin, *et al.*, *Exp. Gerontol.*, 2012, **47**, 179.
80. S. J. Zhao, C. N. Guo, M. Q. Wang, W. J. Chen and Y. B. Zhao, *Cytokine*, 2012, **57**, 221.
81. R. E. Marioni, M. C. Stewart, G. D. Murray, I. J. Deary and F. G. Fawkes, *et al.*, *Psychosom. Med.*, 2009, **71**, 901.
82. N. A. Dennis, S. M. Hayes, S. E. Prince, D. J. Madded, S. A. Huettel and R. Cabeza, *J. Exp. Psychol.: Learn. Mem. Cogn.*, 2008, **34**, 791.
83. J. A. Dumas and P. A. Newhouse, *Pharmacol. Biochem. Behav.*, 2011, **99**, 254.

Parkinson's Disease: Involvement of Iron and Oxidative Stress

DAVID T. DEXTER

Parkinson's Disease Research Group, Centre for Neuroinflammation &
Neurodegeneration, Division of Brain Sciences, Faculty of Medicine,
Imperial College London, Hammersmith Hospital Campus, Du Cane Road,
London W12 0NN, UK
Email: d.dexter@imperial.ac.uk

3.1 Introduction

Parkinson's disease (PD) is the second most common neurodegenerative
disorder affecting man after Alzheimer's disease. The prevalence of PD in
industrialized countries is approximately 0.3% of the whole population. The
prevalence rises with age, which is a major risk factor in PD, from 1% in those
over 60 years of age to almost 4% of the population over 80 years of age. The
mean age of onset is approximately 60 years; however, 10% of cases are classified as young onset, occurring between 20 and 50 years of age.[1] PD is more
prevalent in men than women, with some reports suggesting a ratio of almost
3:1;[2] this difference has been attributed to the neuroprotective effects of
estrogen in females.[3] The economic cost of PD to society is high, but gaining
accurate figures is difficult due to differences in patient care between countries.
The annual cost per year in the UK is estimated to be between £449 million and
£3.3 billion, whilst the cost per patient per year in the US is around $10 000

RSC Metallobiology Series No. 1
Mechanisms and Metal Involvement in Neurodegenerative Diseases
Edited by Roberta Ward, David Dexter and Robert Crichton
© The Royal Society of Chemistry 2013
Published by the Royal Society of Chemistry, www.rsc.org

with a total yearly economic burden of \$23 billion. The greatest proportion of economic costs comes in the latter stages of PD, when a significant proportion of patients enter inpatient care in hospitals and nursing homes, whilst the economic burden coming from medication is substantially lower.[4]

3.2 Clinical Symptomology

3.2.1 Motor Symptoms

The motor features of PD are often used to clinically define the condition; however, some of the non-motor features of the disease often pre-date the occurrence of the motor features. The detection of at least two of the four major motor symptoms are considered cardinal in clinical diagnosis of PD: tremor, bradykinesia, rigidity and postural instability. *Tremor* is the most well known and most apparent symptom of PD. Although it is the most common feature, about 30% of subjects do not have tremor at disease onset, but most patients develop it as the disease progresses. Tremor manifests when the limbs are at rest and disappears with voluntary movement and during sleep. *Bradykinesia* (slowness of movement) not only affects the execution of a movement but also its planning and initiation. The performance of complex sequential or simultaneous movements is particularly affected and this influences daily living skills, *e.g.* getting dressed. *Rigidity* is stiffness and resistance to limb movement due to increased muscle tone resulting from excessive and continuous contraction of muscles. Rigidity is often associated with joint pain, frequently in the neck and shoulders, which is often the initial manifestations of the disease. *Postural instability* is more associated with the later stages of the disease, culminating in poor balance, frequent falls and consequently higher risk of bone fractures. The number of falls is directly related to the disease severity, with approximately 40% of PD subjects experiencing falls.[5] The motor features have a unilateral onset, but as the disease progresses, both sides of the body become affected.

Apart from the classical PD clinical symptoms, other motor features are apparent in a significant proportions of patients, such as gait and posture changes classically manifested in festination (rapid shuffling steps with a forward-flexed posture when walking), speech and swallowing difficulties, mask-like facial expression, micrographia, *etc.*[5]

3.2.2 Non-motor Symptoms

Recently, PD has increasingly become recognized as a more complex illness encompassing both motor and non-motor symptoms (NMS), *e.g.* cognitive decline, sleep disturbances, depression, autonomic dysfunction, *etc.*, with NMS having a major impact on the clinical course of the disease, the biggest determinant of increasing disability, the largest impact on patient quality of life, the largest impact on care costs and dictating whether a patient enters residential care.[6] NMS occur in up to 90% of PD patients, the frequency of NMS increases with PD disease severity, and patients with cognitive impairment have more NMS than those without.[6] There is increasing evidence that some NMS dysfunction, *e.g.* olfactory deficits, constipation, rapid-eye movement (REM), sleep behaviour disorder, depression, *etc.*, antedates clinical PD motor symptoms by many years and may thus be a target for identification of at-risk populations. Current dopamine replacement strategies (see subsequent section) for treating PD are effective against the motor features of PD but are ineffective at addressing the NMS.

3.3 Genetics

In the past, genetics was not considered to have any involvement in the aetiology of PD, due to negative results from studies in twins. However, the discovery of families where PD is a common occurrence and advances in genetics have revealed a number of autosomal dominant and autosomal recessive gene mutations which cause rare familial forms of PD.[7] Gene mutations can be further subdivided in gene mutations which cause either a gain or loss of molecular function (see Table 3.1). Gain of function autosomal dominant mutations in PD include the first gene mutation detected PD, which is in the α-synuclein gene, α-synuclein being the major component of Lewy bodies and Lewy neuritis, which constitute one of the main pathological hallmarks of PD. Considering that α-synuclein association with PD has been known for many years, we still do not know its full function, although it has been demonstrated to be an integral part of neurotransmitter vesicular trafficking in synaptic function. Mutations in the leucine-rich repeat kinase 2 (LRRK2), a mixed linage-like kinase found in membrane-bound structures within cells, forms the most common form of familial PD.[8,9] On the other hand, autosomal loss of function mutations include those in the ubiquitin E3 ligase, parkin, the mitochondrial phosphatase and tensin homologue (PTEN) inducible kinase 1, PINK1, and the rarest mutation in the redox-sensitive

Table 3.1 Common gene mutations causing familial Parkinson's disease.

Gene	Locus/disease	Mode of inheritance	Age of onset (years)
SNCA	PARK1/4	Autosomal dominant	20–85
LRRK2	PARK8	Autosomal dominant	32–79
GRN	FTDP-17[a]	Autosomal dominant	45–83
MAPT	FTDP-17[a]	Autosomal dominant	25–76
DCTN1	Perry syndrome	Autosomal dominant	35–61
PRKN	PARK2	Autosomal recessive	16–72
PINK1	PARK6	Autosomal recessive	20–40
DJ1	PARK7	Autosomal recessive	20–40
FBXO2	PARK15/PPS[b]	Autosomal recessive	10–19
NR4A2/NURR1	–	Unknown	45–67
POLG	–	Unknown	20–26

[a]FTDP-17: frontotemporal dementia with Parkinsonism linked to chromosome 17.
[b]PPS: palido-pyramidal syndrome.

Table 3.2 GWAS risk genes for Parkinson's disease.

Locus	Single nucleotide polymorphism	Protein
MAPT	rs2942168	Microtubule-associated protein tau
SNCA	rs356221	α-Synuclein
BST1	rs4698412	Bone marrow stromal cell antigen 1
LRRK2	rs1491942	Leucine-rich repeat kinase-2
GAK	rs11248051	Cyclin-G-associated kinase
HLA-DRB5	rs3129882	Major histocompatability complex, class II, DR beta5
ACMSD	rs10928513	Aminocarboxymuconate semialdehyde decarboxylase
STK39	rs2102808	Serine threonine kinase 39
HIP1R	rs10847864	Huntingtin interacting protein 1 related
MCCC1/LAMP3	rs11711441	Lysosomal-associated membrane protein 3
SYT11	rs34372695	Synaptotagmin-11
GPNMB	rs156429	Glycoprotein (transmembrane) nmb
STX1B	rs4889603	Syntaxin 1B
FGF20	rs591323	Fibroblast growth factor 20

chaperone DJ-1.[8,9] Although such cases only account for approximately 10% of all diagnosed cases of PD and that often they have distinct clinical and pathological phenotypes, there is sufficient overlap with idiopathic PD that investigations into such gene mutations have revealed important clues as to the molecular mechanisms that underlie the disease process in PD. Indeed, many of molecular mechanism that cause neurodegeneration in the familial PD overlap with the disease mechanisms discovered in idiopathic PD, such as oxidative stress, mitochondrial dysfunction and altered protein handling (see below).[7]

Furthermore, a combination of advances in genetic analysis techniques, the ability to genotype cost effectively and the formation of large patient sample consortia, *e.g.* the International PD Genomics Consortium (IPDGC), have facilitated highly powered genome-wide association studies (GWAS) in idiopathic PD which have revealed some 14 "risk gene" loci for PD (see Table 3.2).[10,11] Whilst the presence of such risk gene loci is not a predictor for the development of PD, they do highlight the fact that PD is a complex disease involving both genetic risk and environmental factors in its aetiology.

3.4 Altered Protein Handling and Neuropathology Consequences

The neuropathological diagnosis of PD requires the identification of two key pathological features, firstly marked dopaminergic neuronal loss in the substantia nigra pars compacta (SNc) and secondly the presence of neuronal inclusions, consisting of misfolded α-synuclein both in the cell bodies as Lewy bodies and in the neuronal process as Lewy neurites.[12] Despite the fact that Friedrich Lewy in 1912 first described Lewy bodies as pale-staining globules, consisting of a dense core surrounded by a pale-staining halo of radiating filaments, we still do not know whether they are protective or neurotoxic. What is known is that the loss of dopaminergic neurons from the nigrostriatal system and the subsequent loss of striatal dopamine precipitate the motor features of PD. However, it is now known that the loss of dopaminergic function also contributes to some of the NMS features of PD, such as sleep disorders, autonomic dysfunction and neuroendocrine problems. In addition to the SNc, other non-dopaminergic nuclei are affected, including the locus ceruleus, reticular formation of the brain stem, the amygdale, raphe nucleus, basal nucleus of the Meynert, dorsal motor nucleus of the vagus, and the hippocampus. The presence of Lewy bodies in the sympathetic ganglia that innervate the heart and sympathetic autonomic denervation of nerves supplying the heart are associated with PD orthostatic hypotension. Constipation is a very common NMS feature in PD, often predating the motor deficits, and this gastrointestinal tract dysfunction is associated with neuronal loss and the presence of Lewy bodies in the dorsal motor nucleus of the vagus, which provides parasympathetic innervations to the visceral organs including the intestine and stomach. Another NMS feature of PD which predates the motor features is hyposmia or loss of smell, which is associated with the presence of Lewy bodies and Lewy neurites in the olfactory bulb and brain centres such as the amygdala and perirhinal nucleus which are associated with olfaction. Additionally, a history of anxiety and depression or both may precede the clinical onset of PD. Whilst the anatomical origins of depression are not fully known, neuronal loss and loss of the neurotransmitters noradrenaline and serotonin in brain areas associated with depression, *e.g.* locus ceruleus and raphe, are a well-recognized feature of PD.

Finally, cognitive decline and dementia are a common late-stage manifestation of PD, with around 85% of patients showing some degree of cognitive decline in the later years of PD. However, there is tremendous variability in the involvement of cognitive decline between patients, with some patients only developing mild cognitive decline after many years of PD whilst other PD patients may develop marked cognitive problems earlier on in the disease course. The pathological causes of cognitive decline or full blown dementia in PD are complex and involve contributions from pathology in monoaminergic and cholinergic nuclei that have cortical projections, the co-existence of Alzheimer's pathology and the presence of diffuse cortical Lewy bodies. Neuronal cell loss in the basal nucleus and cholinergic deficits correlate with the cognitive decline observed in PD subjects that do not have co-existing Alzheimer-type pathology and cortical Lewy bodies. However, there is only a poor response to cholinergic therapies such as acetylcholinesterase inhibitors in PD, suggesting that the cholinergic deficits are not the primary factor in cognitive decline observed in PD.

Whilst a sparse distribution of cortical Lewy bodies are a common feature in most PD cases, in PD with dementia (PDD), cortical Lewy bodies are more widespread and numerous, especially in the limbic lobe. The occurrence of Alzheimer-type amyloid deposits is also a common feature in the brains of

PDD cases. There is also growing evidence pointing to a synergistic interaction between amyloid and α-synuclein pathologies, where there is a significant correlation between cortical amyloid load and limbic cortical α-synuclein deposition. At the other end of the spectrum, if dementia is present at the same time or occurs with one year of the diagnosis of PD, the subject is diagnosed as having dementia with Lewy bodies (DLB), where there is marked neuropathology reflecting both the hallmarks of PD and Alzheimer's disease.

What has become clear over the last decade is that PD pathology does not start in the SNc; indeed, mirroring the occurrence of NMS which precede the motor symptoms of PD, the deposition of α-synuclein and the presence of Lewy bodies is thought to originate in the olfactory bulb and lower brain stem from where it spreads, in stages, to involve the midbrain, eventually spreading to the cortical areas. Braak and colleagues in 2003[13] first proposed a staging of the caudo-rostral spread of α-synuclein pathology and Lewy body deposition in PD. In this hypothesis, Braak proposed that α-synuclein deposition begins in the dorsal motor nucleus of the vagus (stage 1), from where it is thought to proceed in an upward direction *via* the pons (stage 2) to the midbrain (stage 3). From the midbrain the pathology spreads to the basal prosencaphalon and mesocortex (stage 4), finally reaching the neocortex (stages 5 and 6). It is only at stage 3, after extensive loss of dopaminergic neurons in the SNc, that motor features of PD become clinically apparent (for more details, see Table 3.3).[13] However, a number of researchers have disputed the fact that all PD cases exactly follow this formal pattern of spread of PD pathology.[14] Despite this dispute it is now widely accepted that spreading pathology does occur, possibly by differing routes, in PD. This raises the important concept that PD pathology may be propagated from one neuron to another. Exactly how this occurs is not known, but altered α-synuclein released by an affected neuron which is then taken up by an adjacent unaffected neuron, or direct transfer between neurons, could act as a "seed" in a prion-like mechanism to perpetuate the cycle of α-synuclein misfolding and the spread of α-synuclein pathology.[15] Whilst this hypothesis is controversial, some groups have demonstrated such a seeding mechanism in cell culture, animal models and in foetal mesencephalic cells transplanted into the PD striatum.[15] If such a hypothesis is correct, it not only has a major impact on the molecular understanding of PD but also on many other neurodegenerative diseases where altered proteins are involved, and opens up major avenues for drug discovery.

Table 3.3 Braak α-synuclein/Lewy body staging for Parkinson's disease.

Stage	Brain areas affected	Clinical phase of Parkinson's disease
1–2	Dorsal motor nucleus of vagus, raphe nucleus, locus coeruleus	Presymptomatic phase
3	Midbrain: substantia nigra	Presymptomatic phase
4	Amydala, nucleus of Meynert, hippocampus	Symptomatic phase
5–6	Cingulate cortex, temporal cortex, frontal cortex, parietal cortex, occipital cortex	Symptomatic phase

Since Lewy bodies contain a variety of altered proteins, the most common being α-synuclein, this has led to the idea that the catabolism of unwanted, damaged or mutated proteins might be disrupted in PD, leading to cellular aggregation and neuronal death.[16–18] Indeed, several studies have implicated defects in protein handling in both the ubiquitin–proteasome system (UPS) (see Figure 3.1) and lysosomes in the pathogenesis of PD. The potential involvement of the UPS was supported by the uncovering of mutations in parkin and UCH-L1 (yielding two familial forms of PD) which function as an ubiquitin–protein ligase and in ubiquitin recycling, respectively, as components of the UPS. Furthermore, examination of the PD 26S proteasome has revealed selective changes in its catalytic activity and composition in the SNc, which could result in impaired degradation of proteins.[19–21] In cell culture and following direct intra-cerebral injection, proteasomal inhibitors such as lacta-cystin and PSI have been shown to selectively destroy dopaminergic neurones, again supporting the concept that altered protein handling is neurotoxic.[19,20,22]

Exactly how proteasomal inhibition leads to cell death is not known, but a cascade of events involving increase in oxidative and nitrative stress and damage and alterations in mitochondrial function have been detected.[23] Indeed, there is a close association between mitochondrial electron transport activity and the regulatory caps of the 26S proteasome which are ATPases.

Additionally, recent attention has turned to lysosomes and autophagy (macro-autophagy, micro-autophagy and chaperone mediated autophagy) since autophagy-related proteins and autophagosomes are present in Lewy bodies.[24] Indeed, the expression of the chaperone-mediated autophagy proteins LAMP2A and hsc70 is reduced in the PD SNc.[25] Furthermore, silencing of LAMP2A activity in a dopaminergic cell line reduced chaperone-mediated autophagy and increased the half-life of α-synuclein. Altered handling of α-synuclein can also be induced by disruption/mutation of the lysosomal membrane protein ATP13A2 that is affected in one form of familial PD.[26] The

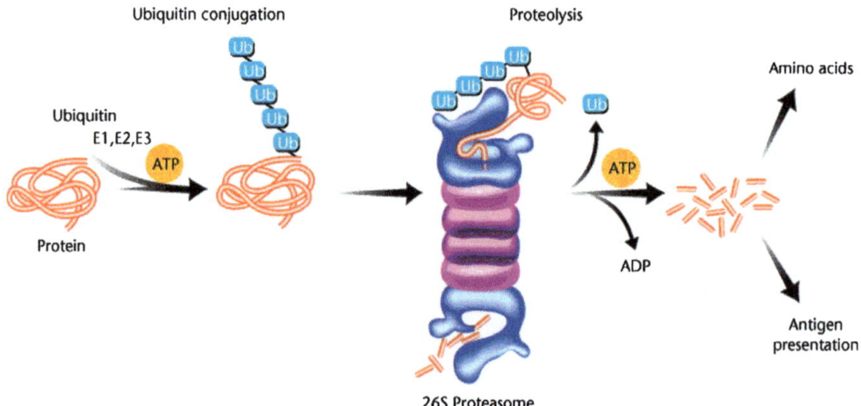

Figure 3.1 Ubiquitination of defective proteins and subsequent breakdown by the proteasome.

potential involvement of altered lysosomal function in PD has also come from toxin-based PD animal models. Administration of the selective dopaminergic neurotoxin 1-methyl-4-phenyl-1,2,3,6-tetrahydropyridine (MPTP) and 6-hydroxydopamine (6-OHDA) caused a depletion of lysosomes in dopaminergic cells secondary to the onset of oxidative stress that led to the accumulation of autophagosomes and cellular degeneration.[27–30] The toxicity of MPTP was attenuated by pharmacological and genetic treatments that enhanced lysosomal formation.

Evidence that lysosomal dysfunction may be involved in neuronal loss in PD has also come from another disease, namely Gaucher disease, a lysosomal storage disorder, due to mutations in the glucocebrosidase gene (GBA). Gaucher disease results in a deficiency of the lysosomal enzyme GCase that catalyses the metabolism of the sphingolipid glucosylceramide to ceramide. Gaucher mutations lead to a 20- to 30-fold increase in risk of developing PD and 5–10% of patients with PD have a GBA mutation.[31] Recently, a GCase deficiency was reported in the SNc of patients with PD and GBA mutations, but importantly, also in those with sporadic PD.[32] Since not all individuals with GBA mutations develop PD, the GCase deficiency may lead to increased susceptibility to other factors involved in the pathogenic process, such as α-synuclein accumulation and oxidative stress occurring as a result of altered mitochondrial function.

3.5 Neuroinflammation

Microglia are from the same lineage as macrophages which are found in the periphery, and are the principal mediators of the innate immune system within the central nervous system. In the normal brain, microglia monitor the brain for any sign of tissue damage or infections and can respond quickly to limit damage and promote repair. Additionally, microglia, in their resting state, secrete factors, *e.g.* growth factors, which promote neuronal survival. Upon detection of tissue damage or infection, microglia become activated, causing them to change morphology, adopting an amoeboid shape and retracting their processes. Microglial activation also leads to the initiation of their antigen presentation and phagocytic properties as well as expressing a variety of inflammatory cytokines, chemokines and proteases.[33]

The concept of inflammatory change in the brain in PD started with the description of activated HLA-positive microglia in the SNc[34] and was subsequently confirmed by Mirza and colleagues.[35] Subsequently, alteration in cytokines (IL-1α, Il-1β, TNF-α) were found in the brain and cerebrospinal fluid (CSF) and post-mortem studies showed inducible nitric oxide synthase (iNOS) to be present in activated microglia.[36] The latter is a source of NO that in turn can react with superoxide from glial or neuronal sources to form the highly reactive peroxynitrite. The latter can nitrate proteins and other biomolecules and 3-nitrotyrosine adducts are found in the PD SNc. Furthermore, in Lewy bodies there is evidence for the presence of nitrated α-synuclein.[37] Recently, positron emission tomography (PET) brain imaging studies utilizing the translocator protein (TSPO) ligand PK-11195, which relatively selectively labels activated microglia, have demonstrated increased microglial activation in the pons, basal ganglia, frontal and temporal regions of PD subjects compared to healthy age-matched controls.[38,39] However, there was no correlation between levels of activation and clinical severity of PD, perhaps suggesting that microglia become activated very early on in the disease process and remain activated. This concept is supported by evidence from Parkinsonism induced in humans by the neurotoxin MPTP, where activated microglia are observed in such subjects up to 16 years after being exposed to this short-lived neurotoxin.[40] Furthermore, microglial activation is observed in the main toxin-based PD animal models 6-OHDA and MPTP, and such activation is observed rapidly after toxin administration and persists for many days, but, importantly, occurs before any neurodegeneration occurs.[41,42] Indeed, activated microglia can be seen surrounding healthy dopaminergic neurons in such models, thus giving support to the concept that "bystander damage" of healthy neurons by activated microglia may be a key feature in PD.[41] What actually activates the microglia is still not known, but there is strong evidence that distress signals from traumatized neurons, *e.g.* matrix metalloproteins (MMPs), alter α-synuclein being released from neurons and neuromelanin being released from degenerating neurons, all potent activators of microglia.

Adding support to the potential damaging effects of microglial activation in PD are the numerous animal studies demonstrating neuroprotection against 6-OHDA or MPTP toxicity by drugs which inhibit microglial activation, *e.g.* iNOS inhibitors, minocycline, peroxisome proliferator-activated receptor gamma (PPAR-γ) agonists, *etc.* (see Figure 3.2).[43–45] Minocycline is a broad-spectrum tetracycline antibiotic that has been shown recently to have anti-inflammatory actions with neuroprotective properties in a variety of neurodegenerative disease models and has entered into Phase III clinical trials in PD. PPAR is a nuclear hormone receptor that was first reported to regulate glucose and lipid metabolism and PPAR-γ agonists such as pioglitazone are

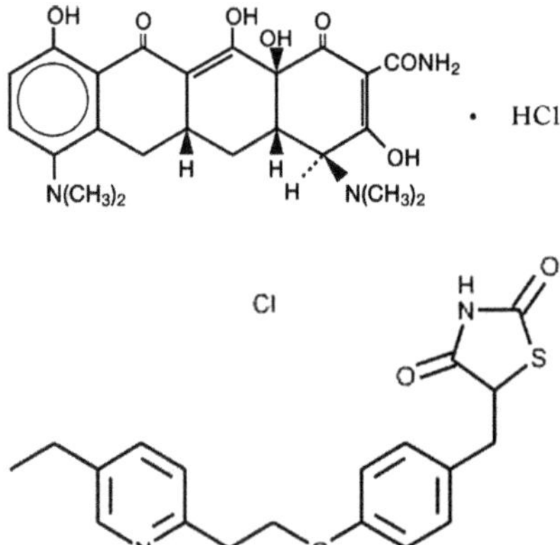

Figure 3.2 Chemical structures of (*top*) minocycline and (*bottom*) pioglitazone.

used in the treatment of diabetes. Subsequently, pioglitazone has been shown to have anti-inflammatory[46] and MMP inhibitor properties.[45,47] Strong neuroprotective findings with PPAR-γ agonists in rodent and primate models of PD have led to PPAR-γ agonists being utilized in Phase II clinical trials in PD.

However, it has been the use of lipopolysaccharides (LPS) both in *in vitro* cellular models and in *in vivo* studies that has provided some compelling evidence that glial activation has a role to play in the PD process.[48–50] These have shown that dopaminergic neuronal loss can occur as a direct consequence of microglial activation that is accompanied by increased cytokine formation, increased production of reactive oxygen and nitrogen species and a decrease in the secretion of trophic factors responsible for the normal maintenance of neuronal viability. A potential important finding is that damage following LPS application to SNc is accompanied by the presence of monocytes, suggesting blood–brain barrier (BBB) damage and permeability (see below). Indeed, altered gene and protein expression for the adhesion molecule ICAM-1 has been shown in PD and in models of the illness.[51]

Microglia activation and inflammatory change was thought to be a consequence of neuronal destruction, but there is evidence for a more general systemic inflammatory reaction in PD suggesting it to be a primary cause of neuronal loss in some cases. In addition, peripheral inflammation may enhance the adverse effects of inflammatory change occurring in the substantia nigra.[52] This point was brought into focus by recent GWAS investigations that showed HLA to be a risk factor for the occurrence of PD.[7]

Additionally, there appears to be some cross-talk between the CNS innate inflammatory response and the peripheral immune system. Microarray gene expression studies have identified modified neuro-immune signalling in peripheral blood mono-nucleated cells (PMCs) from PD patients. Changes in neuroinflammatory markers in PD have more recently been detected by studies demonstrating increased concentrations of interleukin-2 (IL-2),[53] tumour necrosis factor-alpha (TNF-α), IL-6,[54] osteopontin and RANTES/chemokine (C-C motif) ligand 5[55] in the serum of PD subjects. In particular, significant higher serum RANTES levels, a chemokine produced by activated microglia, was not only higher in PD *versus* normal subjects but significantly correlated with unified Parkinson's disease rating scale (UPDRS) scores in the PD.[55] Such findings have supported the concept that a selective blood biomarker may be developed which may be used to refine the clinical diagnosis of PD. Furthermore, there is evidence of an involvement of the adaptive immune system in PD with both CD8$^+$ and CD4$^+$ T-cell lymphocytes, not normally found in the brain, being found in high concentrations in the PD SNc.[34] However, the presence of lymphocytes in the PD CNS may be secondary to the actions of activated microglia which release cytokines that increase the permeability of the BBB and can cause the infiltration of lymphocytes. Furthermore, a high infiltration of CD8$^+$ and CD4$^+$ T-cells is also a feature of the MPTP mouse model of PD,[56] and neuroprotection is associated with dexamethasone administration, which reduces T-cell infiltration, suggesting T-cells may be involved in the neurodegenerative process.[57]

Such data suggest that one way of attempting to reduce cell loss in PD may be through reducing inflammatory change, and certainly this seems to work in experimental systems. This may be relevant to both the initiation and progression of neuronal destruction irrespective of whether this starts at the level of neurones or glia. The importance of the glial activation in PD might lie in its long duration. Conversely, this long-term activation may prevent microglia from performing their normal neuronal support role through the release of growth factors, *etc.*, which are performed in their resting state. While the emphasis on the role of glia has centred on activated microglia, it should be borne in mind that astrocytosis also occurs in PD and astrocytes may also play a significant role in the sequence of events that leads to cell death.

3.6 Role of Iron and Oxidative Stress in Parkinson's Disease

Iron plays a vital role in many physiological functions, such as DNA synthesis, mitochondrial respiration and oxygen transport. Neuronally, iron is involved in myelination and neurotransmission and is the most abundant brain metal. Additionally, iron acts as a cofactor for tyrosine hydroxylase (TH), the enzyme at the rate-limiting step in the synthesis of dopamine. Whilst iron plays a key role in these physiological functions, excess iron can be toxic through oxidative stress.

We were the first research group to demonstrate that the neurodegenerative process in PD is associated with elevated iron levels in the SNc.[58–60] Since then, other post-mortem studies, MRI and transcranial ultrasound studies have verified these observations.[61–63] Electron probe X-ray microanalysis has been utilized to verify that individual dopaminergic neurons in the PD SN have raised iron levels.[64] Furthermore, iron accumulation has also been linked to other neurological disorders with Parkinsonian features.[65] Immunohisto-chemistry studies show a 60% reduction in iron–transferrin receptor binding in PD as well as a decrease in transferrin binding sites.[66,67] We have also demonstrated that, in PD brains, excessive iron is not mirrored by an increase in ferritin levels;[68] normally, brain H-ferritin levels parallel the increased iron accumulation which occurs with age.[69] Furthermore, in dopamine neurons, iron stored within the ferritin core can be reduced readily by products of dopamine oxidation such as 6-OHDA, adding a further level of vulnerability to this neuronal population.[70] In an aged individual, the ferritin can become heavily burdened with the excess iron, which accumulates with age. Within lysosomes, ferritin may become degraded, releasing the bound iron and further increasing the level of reactive iron.[71] Recent studies have also implicated iron in promoting α-synuclein aggregate formation;[72,73] hence the accumulation of iron could contribute to other pathological process in PD apart from stimu-lating oxidative stress (see Figure 3.3). Additionally, iron has been shown to accumulate in the SNc following 6-OHDA, MPTP or the proteasome inhibitor lactacystein,[74–76] thus providing further support for a role for iron in the

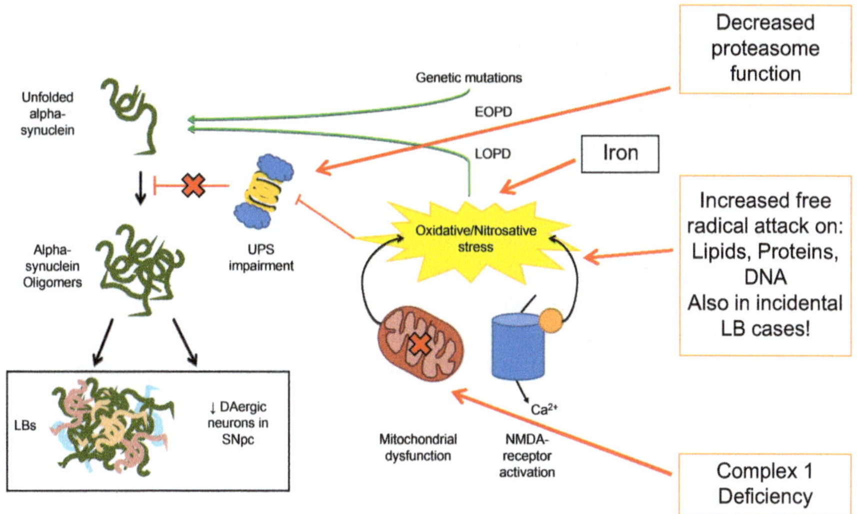

Figure 3.3 Involvement of oxidative stress in the neurodegenerative process in Parkinson's disease.

neurodegenerative process. While direct evidence for increased free radical generation due to iron accumulation is almost impossible to detect, there is ample indirect evidence of oxidative stress which has withstood many years of research.[77–80] This includes enhanced lipid peroxidation (malondialdehyde, lipid hydroperoxides, 4-hydroxynonenal and advanced glycation end products), protein oxidation (protein carbonyls) and DNA oxidation (8-hydroxyguanosine) in the SNc in PD (see Figure 3.3). More recently, the finding of a negative correlation between plasma urate levels and disease progression in PD has raised the spectrum of altered antioxidant activity that reflects the decrease previously reported in reduced glutathione levels (along with changes in catalase and glutathione peroxidase).[79,81,82]

3.6.1 Altered Mitochondrial Function in PD

The involvement of mitochondrial defects in cell death in PD has been recognized for many years, and the recent discovery of their role in familial PD has revealed they are part of a unifying concept of how neuronal loss occurs in both sporadic and inherited disease. The discovery of the neurotoxicity for dopaminergic neurons of MPTP through its metabolite MPP^+ identified a role for the inhibition of complex I in PD pathogenesis.[83] Very quickly, an inhibition of complex I was shown to be present in the PD SNc that was tissue and disease specific (see Figure 3.3).[84] The construction of cybrids using mtDNA from patients with PD has clearly demonstrated the encoding of the complex I defect.[85] Importantly, only about 30% of patients with PD have a clear complex I defect compared to controls, suggesting, perhaps, that they form an important sub-group within the disease that offer a specific target for interference with the disease process.

More recently, genetic investigations in familial PD have also revealed a role of mitochondria in this class of PD patients (see Figure 3.4). Notably, mutations in α-synuclein, parkin, PINK1, DJ-1 and, perhaps, LRRK2 have been associated with altered mitochondrial function.[86,87] Such gene mutations in PD lead to altered protein localization in mitochondria, mitochondrial structured abnormalities and function and to a decrease in complex I activity and assembly. Loss of function of, notably, DJ-1 but also parkin and PINK1 decreases mitochondrial protection against oxidative stress, which in turn increases mitochondrial dysfunction. Another important role for parkin and PINK1 is in the turnover of mitochondria by autophagy, specifically mitophagy, where they act in tandem to regulate this process. This may be critically important in PD where autophagy appears impaired, so reducing the cells ability to remove damaged mitochondria.

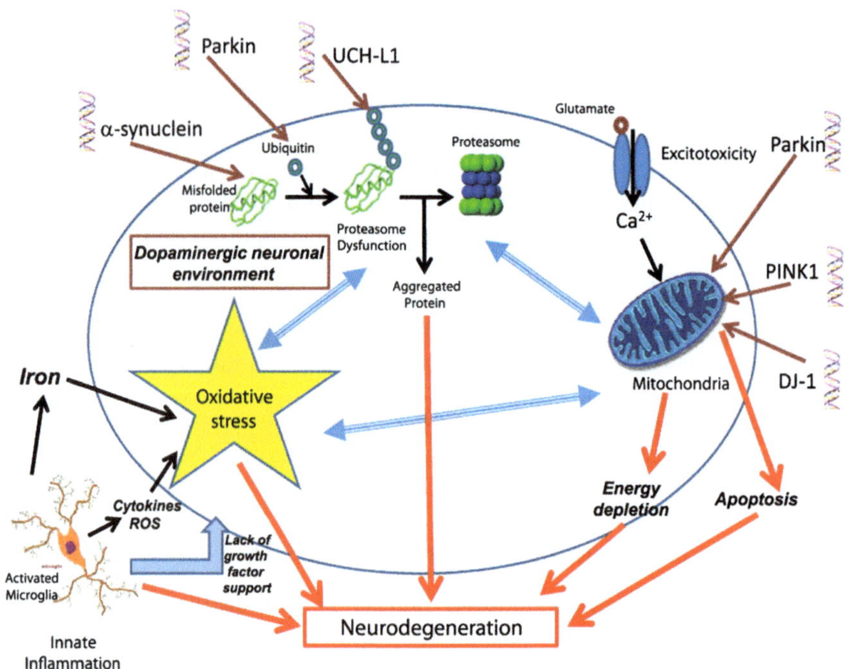

Figure 3.4 Overlap of currently known neurodegenerative disease mechanisms in idiopathic and familial forms of PD. The figure represents the disease mechanisms within the dopaminergic neurons. *Blue double-headed arrows* indicate molecular mechanisms which may not only be toxic in their own right but may also influence other disease mechanisms known to be a feature of cell death mechanisms in PD. *Double helix structures* represent some of the common gene mutations found in familial PD and *brown arrows* indicate where the resultant altered proteins interfere with cell function and overlap known mechanisms of cell death in idiopathic PD.

3.7 Therapeutic Strategies

3.7.1 Current Approaches to Treatment

The treatment of PD has not changed substantially in the past 30 years, with dopamine replacement therapy employing dopamine agonists and L-DOPA as the mainstay, supported by the use of a series of enzyme inhibitors responsible for the breakdown of L-DOPA or dopamine, namely peripheral decarboxylase inhibitors, catechol-*O*-methyltransferase (COMT) inhibitors and monoamine oxidase-B (MAO-B) inhibitors (see Figure 3.5). Outside of the dopaminergic arena, only anti-cholinergics and the weak *N*-methyl-D-aspartate (NMDA) receptor antagonist, amantadine, have had any use. However, all of these treatment strategies are symptomatic approaches to treating the motor deficits of PD, with little effect on non-motor symptoms and no proven effect on disease progression.

There have been several attempts to determine whether current therapy has any effect on the disease process. Two dopamine agonists, ropinirole and pramipexole, show neuroprotective actions in preclinical animal models of PD loss, but in clinical trials (REAL-PET, CALM-PD) there was no slowing of progression of motor symptoms.[88,89] Even in mild PD, the early use of pramipexole showed no advantage over the later introduction of the drug (PROUD study).[90]

L-DOPA has had a chequered career as both a potential reason for acceleration of the cell death in PD and for its slowing. For many years, the ability of L-DOPA to generate free radicals and to kill dopaminergic cells in culture tainted the drug with a neurotoxic label, but overall, based on a range of *in vivo* studies in animal, post-mortem studies in man and in clinical experience, this does not appear to be true.[91,92] Rather, from the ELLDOPA clinical trial among others, early use of L-DOPA appears to slow disease progression based

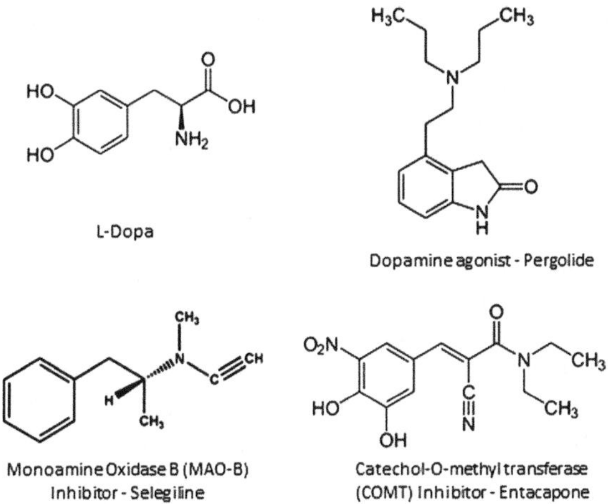

Figure 3.5 Chemical structures of typical anti-Parkinson's medication.

on clinical rating scales but not when using imaging endpoints.[93] However, it has been the MAO inhibitors selegiline and rasagiline that have attracted most attention as potential disease modifiers. The principal actions of such drugs are to prevent the breakdown of dopamine *via* MAO, thus preserving what dopamine is in the brain and reducing dramatically the dose of L-DOPA required to control PD. However, both drugs also are associated with a range of actions that encompass anti-apoptotic actions, antioxidant effects, anti-glutamatergic effects and neurotrophic actions, among others, that underlie a belief in their ability to alter the rate of loss of dopaminergic neurones.[94,95] Both selegiline and rasagiline have been examined in detail in clinical trials to assess whether disease modification occurs in man. The DATATOP study initially suggested that selegiline treatment could delay the need for the introduction of L-DOPA in early PD, but this now appears largely due to a symptomatic effect of the drug, although open label extension studies in other clinical trials still suggest that disease progression is slower in those patients with PD that receive early selegiline treatment.[96] So too with rasagiline, with the results of the TEMPO and ADAGIO studies suggesting that early use had a positive effect on clinical scores compared to later introduction of the drug, most notably in those patients with the highest degree of motor disability.[88,97] However, so far, none of the studies undertaken have convinced the regulatory authorities to label either selegiline or rasagiline as disease modifying.

3.7.2 Novel Therapeutic Approaches

3.7.2.1 *Modulation of Glutamate Activity*

This stems from overactivity of the glutamatergic subthalamic nucleus (STN) pathways which innervate the SNc and the internal segment of the globus pallidus.[98,99] There is also evidence of altered glutamatergic input from the cortico-striatal pathway. Not only can the overactivity of such pathways account for a large proportion of the symptoms of PD, the release of excessive amounts of glutamate onto the dopaminergic neurons in the SNc could exacerbate the neurodegenerative process by causing excitotoxicity (see Figure 3.6). A scenario that has been proposed is that as PD develops with the onset of neuronal loss in the SNc, the increased activity of the glutamatergic input from the STN acts to amplify cell death in the SNc. Indeed, lesions of the STN have been shown to reduce the loss of dopaminergic neurones normally induced by the injection of 6-OHDA into the SNc.[100] Indeed, deep brain stimulation (DBS) has been utilized to correct the overactivity of the STN, producing clinical benefit in PD. Recent studies utilizing DBS in PD subjects has been associated with a lack of motor symptom progression in PD.[101] Assuming that DBS would also decrease the overactivity of the STN pathways innovating the SNc, this does add some support to the concept that excito-toxicity does have a role to play in PD. Hence, drugs which modify glutamate activity could not only affect the clinical symptoms of PD but could also slow the disease down by limiting excitotoxicity. Although inotropic glutamate

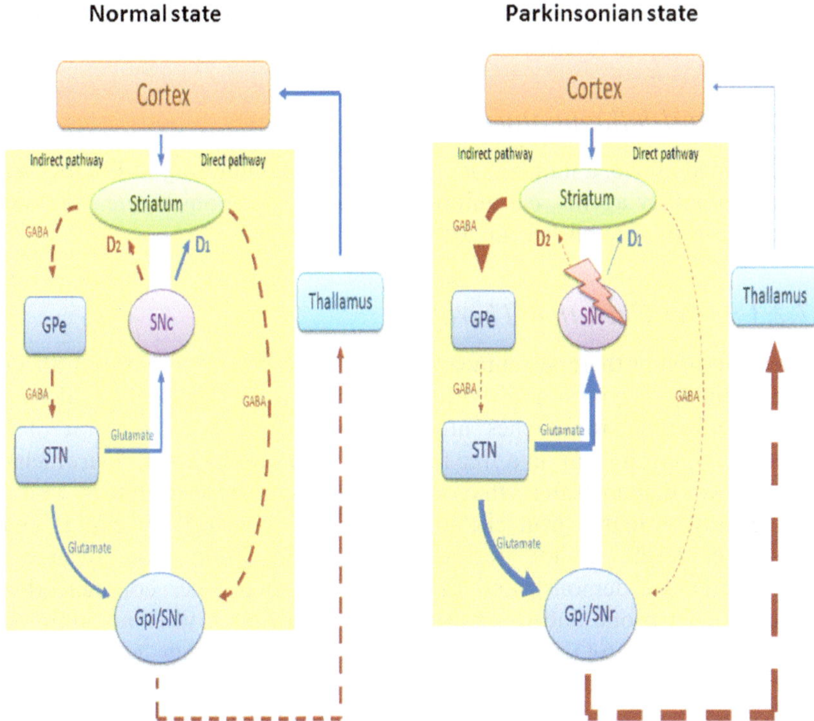

Figure 3.6 Innovation of the basal ganglia in the normal and Parkinson's brain. *Solid blue lines* represent excitatory pathways and *broken red lines* represent inhibitory pathways. The basal ganglia's main input centre is the striatum, which receives motor control signals primarily from the cortex. The main output centre comprises the globus pallidus internus (GPi) and the substantia nigra pars reticulate (SNr) and works to inhibit thalamcortical signalling. The SNc controls basal ganglia activity by modulation of the output centre through the direct and indirect pathways which inhibit or enhance the output centre's signalling, respectively. The loss of dopaminergic neurons in the PD SNc leads to a decrease in activity in the direct pathway and an increase in activity in the indirect pathway, resulting in overactivation of the output centre and the emergence of motor symptoms. In addition, increased STN glutamatergic signalling to the SNc increases nigral cell death through excitotoxic mechanisms.

receptor antagonists initially showed promise in animal models of PD, they had a narrow therapeutic effect in the clinical trial that was associated with marked side effects principally due to the wide distribution of such receptors in the brain.

Recent attention has turned to drugs which act on the metabotropic glutamate receptors (mGluRs) since they have a more selective distribution within the brain, principally in the basal ganglia, and their presence on microglial and astrocytic cells has been proposed to modify the inflammatory response. Studies in animal models have demonstrated both neuroprotective and disease symptom modifying effects (for review, see Duty[102]); however, such drugs have yet to be trialled on PD subjects.

3.7.2.2 Anti-inflammatory Drugs

As mentioned above, a number of clinical trials are underway in PD looking at anti-inflammatory agents, *e.g.* minocycline, PPAR-γ agonists, *etc.*, the results of which are to be published in 2013.

3.7.2.3 Iron Chelation Therapy

Disease-modifying therapies aimed at removing excess iron, without affecting iron-containing enzymes involved in neurotransmitter function, could be part of the therapeutic approaches utilized to prevent the progression of PD. However, such iron is not in a free form but bound primarily to neuromelanin. In 1994 we demonstrated that novel bidentate iron chelators and the commercially available iron chelator desferrioxamine (DFO) could remove excess iron levels in the ferrocene model of brain iron overload.[103] Subsequently, we demonstrated that DFO and two other commercial iron chelators, deferiprone and desferasirox (see Figure 3.7), were not only neuroprotective in the 6-OHDA models of PD but also they could scavenge hydroxyl radical formation as assessed with *in vivo* microdialysis.[104] Such chelators are extensively utilized to treat peripheral iron overload disorders such as beta thalassaemia, where the patients requires regular blood transfusions but iron accumulates in the body when the red blood cells are removed from the circulation.

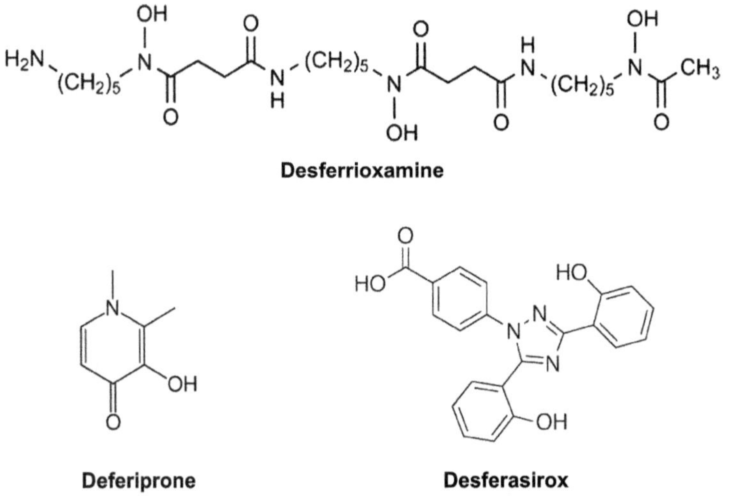

Figure 3.7 Chemical structures of the three commercially available iron chelators, desferrioxamine, deferiprone and desferasirox, currently used to clinically treat beta thalassaemia.

In 2007, deferiprone was used in a pilot clinical trial in Friedreich's ataxia (FA) patients, which occurs due to a gene defect leading to iron accumulation in the mitochondria in the cerebellum, toxicity of which leads to ataxia or inability to control muscle movements. FA is diagnosed in young individuals (14–26 years of age) and has no clinical treatment. In this six-month FA study, deferiprone at 20 or 30 mg kg^{-1} d^{-1} was well tolerated by the patients and resulted in a reduction in brain iron, as indicated by MRI brain imaging, and led to a clinical improvement in patient symptoms.[105] The positive effects with marked side effects of this first clinical use of deferiprone for a neurological disorder stimulated the establishment of two pilot clinical trials with deferiprone, the PAIR-PARK-1 study at University Hospital Lille and the deferiprone PD trial here at Imperial College London. Both trials are due to report in 2013; if results are positive, larger multicentre trials will be organized.

References

1. C. M. Tanner, F. Kamel, G. W. Ross and J. A Hoppin, *et al.*, *Environ. Health Perspect.*, 2011, **119**, 866.
2. A. Schrag, Y. Ben-Shlomo and N. P. Quinn, *Br. Med. J.*, 2000, **321**, 21.
3. G. E. Gillies, H. E. Murray, D. T. Dexter and S. McArthur, *Pharmacol. Biochem. Behav.*, 2004, **78**, 513.
4. L. J. Findley, *Parkinsonism Relat. Disord.*, 2007, **13**, S8.
5. F. Jankovic, *J. Neurol. Neurosurg. Psychiatr.*, 2008, **79**, 368.
6. K. R. Chaudhuri and A. H. Schapira, *Lancet Neurol.*, 2009, **8**, 464.
7. H. Houlden and A. B. Singleton, *Acta Neuropathol.*, 2012, **124**, 325.
8. A. Gupa, V. L. Dawson and T. M. Dawson, *Ann. Neurol.*, 2008, **64**, S3.
9. T. Dawson, *Parkinsonism Relat. Disord.*, 2007, **13**, S248.
10. International Parkinson's Disease Genomics Consortium (IPDGC), Wellcome Trust Case Control Consortium 2 (WTCCC2), *PLoS Genet.*, 2011, **7**, e1002142.
11. International Parkinson Disease Genomics Consortium, *Lancet*, 2011, 377, 641.
12. D. W. Dickson, H. Braak, J. E. Duda, C. Duyckaerts, T. Gasser and G. M. Halliday, *et al.*, *Lancet Neurol.*, 2009, **8**, 1150.
13. H. Braak, K. Del Tredici, U. Rub, R. A. de Vos, E. N. Jansen Steur and E. Braak, *Neurobiol. Aging*, 2003, **24**, 197.
14. M. E. Kalaitzakis, M. B. Graeber, S. M. Gentleman and R. K. B. Pearce, *Neuropathol. Appl. Neurobiol.*, 2009, **34**, 284.
15. C. J. R. Dunning, J. F. Reyes, J. A. Steiner and P. Brindin, *Prog. Neurobiol.*, 2012, **97**, 205.
16. B. I. Giasson, J. E. Duda, I. V. Murray, Q. Chen, J. M. Souza, H. I. Hurtig, H. Ischiropoulos, J. Q. Trojanowski and V. M. Lee, *Science*, 2000, **290**, 985.
17. P. Shashidharan, P. F. Good, A. Hsu, D. P. Perl, M. F. Brin, C. W. Olanow and A. Torsin, *Brain Res.*, 2000, **877**, 379.

18. M. G. Spillantini, R. A. Crowther, R. Jakes, M. Hasegawa and M. Goedert, *Proc. Natl. Acad. Sci. U. S. A.*, 1998, **95**, 6469.

19. K. S. McNaught and P. Jenner, *Neurosci. Lett.*, 2001, **297**, 191.

20. K. S. McNaught, P. Shashidharan, D. P. Perl, P. Jenner and C. W. Olanow, *Eur. J. Neurosci.*, 2002, **16**, 2136.

21. G. K. Tofaris, R. Layfield and M. G. Spillantini, *FEBS Lett.*, 2001, **509**, 22.

22. W. Xie, X. Li, C. Li, W. Zhu, J. Jankovic and W. Le, *J. Neurochem.*, 2010, **115**, 188.

23. D. H. Hyun, M. Lee, B. Halliwell and P. Jenner, *J. Neurochem.*, 2003, **86**, 363.

24. A. H. Schapira, *Mt. Sinai J. Med.*, 2011, **78**, 872.

25. L. Alvarez-Erviti, M. C. Rodriguez-Oroz, J. M. Cooper, C. Caballero, I. Ferrer, J. A. Obeso and A. H. Schapira, *Arch. Neurol.*, 2010, **67**, 1464.

26. M. Usenovic, E. Tresse, J. R. Mazzulli, J. P. Taylor and D. Krainc, *J. Neurosci.*, 2012, **32**, 4240.

27. B. Dehay, J. Bove, N. Rodriguez-Muela, C. Perier, A. Recasens, P. Boya and M. Vila, *J. Neurosci.*, 2010, **30**, 12535.

28. B. Dehay, A. Ramirez, M. Martinez-Vicente, C. Perier, M. H. Canron, E. Doudnikoff, A. Vital, M. Vila, C. Klein and E. Bezard, *Proc. Natl. Acad. Sci. U. S. A.*, 2012, **109**, 9611.

29. C. Marin and E. Aguilar, *Neurochem. Int.*, 2011, **58**, 521.

30. M. E. Solesio, S. Saez-Atienzar, J. Jordan and M. F. Galindo, *Toxicol. Sci.*, 2012, **129**, 411.

31. A. Velayati, W. H. Yu and E. Sidransky, *Curr. Neurol. Neurosci. Rep.*, 2010, **10**, 190.

32. M. E. Gegg, D. Burke, S. J. Heales, J. M. Cooper, J. Hardy, N. Wood and A. H. Schapira, *Ann. Neurol.*, 2012, **72**, 455.

33. E. C. Hirsch and S. Hunot, *Lancet Neurol.*, 2009, **8**, 382.

34. P. L. McGeer, S. Itagaki, B. E. Boyes and E. G. McGeer, *Neurology*, 1988, **38**, 1285.

35. B. Mirza, H. Hadberg, P. Thomsen and T. Moos, *Neuroscience*, 2000, **95**, 425.

36. E. C. Hirsch, T. Breidert, E. Rousselet, S. Hunot, A. Hartmann and P. P. Michel, *Ann. N. Y. Acad. Sci.*, 2003, **991**, 214.

37. B. I. Giasson, J. E. Duda, I. V. Murray and Q. Chen, *et al.*, *Science*, 2000, **290**, 985.

38. A. Gerhard, N. Pavese, G. Hotton, F. Turkheimer, M. Es and A. Hammers, *et al.*, *Neurobiol. Dis.*, 2006, **21**, 404.

39. Y. Ouchi, E. Yoshikawa, Y. Sekine, M. Futatsubashi, T. Kanno and T. Ogusu, *et al.*, *Ann. Neurol.*, 2005, **57**, 168.

40. J. W. Langston, L. S. Forno, J. Tetrud, A. G. Reeves, J. A. Kaplan and D. Karluk, *Ann. Neurol.*, 1999, **46**, 598.

41. L. Marinova-Mutafchieva, M. Sadeghian, L. Broom, J. B. Davis, A. D. Medhurst and D. T. Dexter, *J. Neurochem.*, 2009, **110**, 966.

42. G. T. Liberatore, V. Jackson-Lewis, S. Vukosavic, A. S. Mandir, M. Vila and W. G. McAuliffe, *et al.*, *Nat. Med.*, 1999, **5**, 1403.

43. L. Broom, L. Marinova-Mutafchieva, M. Sadeghian, J. B. Davis, A. D. Medhurst and D. T. Dexter, *Free Radical Biol. Med.*, 2011, **50**, 633.

44. Y. He, S. Appel and W. Le, *Brain Res.*, 2001, **909**, 187 .

45. M. Sadeghian, L. Marinova-Mutafchieva, L. Broom, J. B. David, D. Virley, A. D. Medhurst and D. T. Dexter, *J. Neuroimmunol.*, 2012, **246**, 69.

46. R. B. Clark, *J. Leukocyte Biol.*, 2002, **71**, 388.

47. S. Zafiriou, S. R. Stanners, S. Saad, T. S. Polhill, P. Poronnik and C. A. Pollock, *J. Am. Soc. Nephrol.*, 2005, **16**, 638.

48. M. M. Iravani, C. C. Leung, M. Sadeghian, C. O. Haddon, S. Rose and P. Jenner, *Eur. J. Neurosci.*, 2005, **22**, 317.

49. M. M. Iravani, M. Sadeghian, C. C. Leung, P. Jenner and S. Rose, *Neurosci. Lett.*, 2012, **510**, 138.

50. W. G. Kim and R. P. Mohney, *et al.*, *J. Neurosci.*, 2000, **20**, 6309.

51. J. Miklossy, D. D. Doudet, C. Schwab, S. Yu, E. G. McGeer and P. L. McGeer, *Exp. Neurol.*, 2006, **197**, 275.

52. M. C. Hernandez-Romero, M. J. Gado-Cortes, M. Sarmiento, R. M. de Pablos, A. M. Espinosa-Oliva, S. Arguelles and M. J. Bandez, *et al.*, *Neurotoxicology*, 2012, **33**, 347.

53. G. Stypuła, J. Kunert-Radek, H. Stepień, K. Zylińska and M. Pawlikowski, *Neuroimmunomodulation*, 1996, **3**, 131.

54. R. J. Dobbs, A. Charlett, A. G. Purkiss, S. M. Dobbs, C. Weller and D. W. Peterson, *Acta Neurol. Scand.*, 1999, **100**, 34.

55. M. Rentzos, C. Nikolaou, E. Andreadou, G. P. Paraskevas, A. Rombos, M. Zoga, A. Tsoutsou, F. Boufidou, E. Kapaki and D. Vassilopoulos, *Acta Neurol. Scand.*, 2007, **119**, 332.

56. V. Brochard, B. Combadiere, A. Prigent, Y. Laouar, A. Perrin and V. Beray-Barthat, *et al.*, *J. Clin. Invest.*, 2009, **119**, 182.

57. I. Kurkowska-Jastrzebska, A. Wronska, M. Kohutnicka and A. Czlonkowska, *Exp. Neurol.*, 1999, **156**, 50.

58. D. T. Dexter, F. Agid, Y. Agid, F. R. Wells, A. J. Lees, P. Jenner and C. D. Marsden, *Lancet*, 1987, **ii**, 1219.

59. D. T. Dexter, F. Agid, Y. Agid, F. R. Wells, A. J. Lees, P. Jenner and C. D. Marsden, *J. Neurochem.*, 1989, **52**, 1830.

60. K. A. Jellinger, *Mol. Chem. Neuropathol.*, 1991, **14**, 153.

61. M. E. Gotz, K. Double, M. Gerlach, M. B. Youdim and P. Riederer, *Ann. N. Y. Acad. Sci.*, 2004, **1012**, 193.

62. Y. H. Han, J. H. Lee, B. M. Kang, C. W. Mun, S. K. Baik, Y. I. Shin and K. H. Park, *J. Neurol. Sci.*, 2013, **325**, 29; doi: 10.1016/j.jns.2012.11.009.

63. U. Walter, *Int. Rev. Neurobiol.*, 2010, **90**, 166.

64. A. E. Oakley, J. F. Collingwood and G. Dobson, *Neurology*, 2007, **68**, 1820.

65. D. T. Dexter, A. Carayon, F. Agid, Y. Agid, S. E. Daniel, A. J. Lees, P. Jenner and C. D. Marsden, *Brain*, 1991, **114**, 1953.

66. C. M. Morris, J. M. Candy, S. Omar, C. A. Bloxham and J. A. Edwards, *Neuropathol. Appl. Neurobiol.*, 1994, **20**, 68.
67. B. A. Faucheux, J. Haux, Y. Agid and E. C. Hirsch, *Brain Res.*, 1997, **749**, 170.
68. D. T. Dexter, A. Carayon, F. Agid, Y. Agid and S. E. Daniel, *et al.*, *J. Neurochem.*, 1990, **55**, 16.
69. B. Hallgren and P. Sourander, *J. Neurochem.*, 1958, **3**, 41.
70. H. P. Monterio and C. C. Winterbourn, *Biochem. Pharmacol.*, 1989, **38**, 4177.
71. K. L. Double, M. Matwald, M. Schmittel, P. Riederer and M. Gerlach, *J. Neurochem.*, 1998, **70**, 2492.
72. S. Mandel, G. Maor and M. B. H. Youdim, *J. Mol. Neurosci.*, 2004, **24**, 401.
73. N. Ostrerova-Golts, L. Petrucelli, J. Hardy, J. M. Lee, M. Farer and B. Wolozin, *J. Neurosci.*, 2000, **20**, 6048.
74. N. Song, H. Jiang, J. Wang and J. X. Xie, *J. Neurosci. Res.*, 2007, **18**, 1181.
75. H. Mochizuki, H. Imai, K. Endo, K. Yokomizo, Y. Murata, N. Hattori and Y. Mizuno, *Neurosci Lett.*, 1994, **168**, 251.
76. W. Zhu, W. Xie, T. Pan, P. Xu, M. Fridkin, H. Zheng, J. Jankovic, M. B. Youdim and W. Le, *FASEB J*, 2007, **21**, 3835.
77. P. Jenner, *Ann. Neurol.*, 2003, **53**(suppl. 3), S26.
78. P. Jenner and C. W. Olanow, *Neurology*, 2006, **66**, S24.
79. J. Sian, D. T. Dexter, A. J. Lees, S. Daniel, Y. Agid, F. Javoy-Agid, P. Jenner and C. D. Marsden, *Ann. Neurol.*, 1994, **36**, 348.
80. D. T. Dexter, F. Agid, Y. Agid, F. R. Wells, A. J. Lees, P. Jenner and C. D. Marsden, *J. Neurochem.*, 1989, **52**, 381.
81. L. M. Ambani, M. H. Van Woert and S. Murphy, *Arch. Neurol.*, 1975, **32**, 114.
82. A. Ascherio, P. A. LeWitt, K. Xu, S. Eberly, A. Watts, W. R. Matson and C. Marras, *et al.*, *Arch. Neurol.*, 2009, **66**, 1460.
83. R. R. Ramsay, J. I. Salach and T. P. Singer, *Biochem. Biophys. Res. Commun.*, 1986, **134**, 743.
84. A. H. Schapira, *Lancet Neurol.*, 2008, **7**, 97.
85. M. Gu, J. M. Cooper, J. W. Taanman and A. H. Schapira, *Ann. Neurol.*, 1998, **44**, 177.
86. A. H. Schapira and P. Jenner, *Mov. Disord.*, 2011, **26**, 1049.
87. A. H. Schapira, *Mt. Sinai J. Med.*, 2011, **78**, 872.
88. Parkinson Study Group, *Arch. Neurol.*, 2002, **59**, 1937.
89. A. L. Whone, R. L. Watts, A. J. Stoessl, M. Davis, S. Reske, C. Nahmias and A. E. Lang, *et al.*, *Ann. Neurol.*, 2003, **54**, 93.
90. A. H. Schapira, S. Albrecht, P. Barone, C. L. Comella, M. P. McDermott, Y. Mizuno, W. Poewe, O. Rascol and K. Marek, *Mov. Disord.*, 2010, **25**, 1627.
91. L. Parkkinen, S. S. O'Sullivan, M. Kuoppamaki, C. Collins, C. Kallis, J. L. Holton, D. R. Williams, T. Revesz and A. J. Lees, *Neurology*, 2011, **77**, 1420.

92. C. W. Olanow, Y. Agid, Y. Mizuno, A. Albanese, U. Bonuccelli and P. Damier, *et al.*, *Mov. Disord.*, 2004, **19**, 997.
93. S. Fahn, *J. Neurol.*, 2005, **252**(suppl. 4), IV37.
94. P. Jenner and J. W. Langston, *Mov. Disord.*, 2011, **26**, 2316.
95. S. Mandel, O. Weinreb, T. Amit and M. B. Youdim, *Brain Res. Rev.*, 2005, **48**, 379.
96. S. Palhagen, E. Heinonen, J. Hagglund, T. Kaugesaar, O. Maki-Ikola and R. Palm, *Neurology*, 2006, **66**, 1200.
97. C. W. Olanow, R. A. Hauser, J. Jankovic, W. Langston, A. Lang, W. Poewe, E. Tolosa, F. Stocchi, E. Melamed, E. Eyal and O. Rascol, *Mov. Disord.*, 2008, **23**, 2194.
98. M. Bevan, P. Magill, D. Terman, P. Bolam and C. Wilson, *Trends Neurosci.*, 2002, **25**, 525.
99. F. Steigerwald, M. Potter, J. Herzog, M. Pinsker, F. Kopper, H. Mehdron, G. Deuschl and J. Volkmann, *J. Neurophysiol.*, 2008, **100**, 2515.
100. B. Wallace, K. Ashkan, C. Heise, K. Foote, N. Torres, J. Mitrofanis and A. Benabid, *Brain*, 2007, **130**, 2129.
101. M. Tagliati, C. Martin and R. Alterman, *Int. J. Neurosci.*, 2010, **120**, 717.
102. S. Duty, *CNS Drugs*, 2012, **26**, 1017.
103. R. J. Ward, D. T. Dexter, A. Florence, F. Aouad, R. Hider, P. Jenner and R. R. Crichton, *Biochem. Pharmacol.*, 1995, **48**, 1821.
104. D. T. Dexter, S. A. Statton, C. Whitmore, W. Freinbichler, P. Weinberger, K. F. Tipton, L. Della Corte, R. J. Ward and R. R. Crichton, *J. Neural Transm.*, 2011, **118**, 223.
105. N. Boddaert, K. H. L. Q. Sang, A. Rotig, A. Leroy-Willig, S. Gallet, F. Brunelle, D. Sidi, J. Thalabard, A. Munnich and Z. L. Cabantchik, *Blood*, 2007, **110**, 401.

The Role of Metals in Alzheimer's Disease

NABIL HAJJI,[a] CARLY CALVERT,[b] CRAIG W. RITCHIE[b] AND MAGDALENA SASTRE*[b]

[a] Division of Experimental Medicine and [b] Division of Brain Sciences, Imperial College London, Hammersmith Hospital Campus, Du Cane Road, London W12 0NN, UK
*Email: m.sastre@imperial.ac.uk

4.1 Introduction

Alzheimer's disease (AD) is the most common cause of dementia and the main pathological feature is massive neuronal loss in areas of the brain responsible for memory and learning, such as the cortex and hippocampus. In the brain, three major hallmarks are associated with the processes of this disease: amyloid beta peptide (Aβ) deposition, neurofibrillary tangles (NFTs) of hyperphosphorylated microtubule-associated tau and synaptic loss (Figure 4.1).[1,2] Epidemiological studies have shown an association between heavy metals, such as lead, cadmium and mercury, and AD.[3] While these metals have no biological function with the pathogenesis of AD, others such as iron, copper or zinc fulfil various essential biological functions where any changes in their levels by excess or deficit can lead to deleterious responses and alter cognitive functions.[3]

Insufficient supply of oxygen to the brain is one of the major effects of the heavy metals at both poisoning levels and long-term exposure to low levels, resulting in an anoxia/hypoxic brain (no oxygen or little oxygen) and leading to

RSC Metallobiology Series No. 1
Mechanisms and Metal Involvement in Neurodegenerative Diseases
Edited by Roberta Ward, David Dexter and Robert Crichton

Published by the Royal Society of Chemistry, www.rsc.org

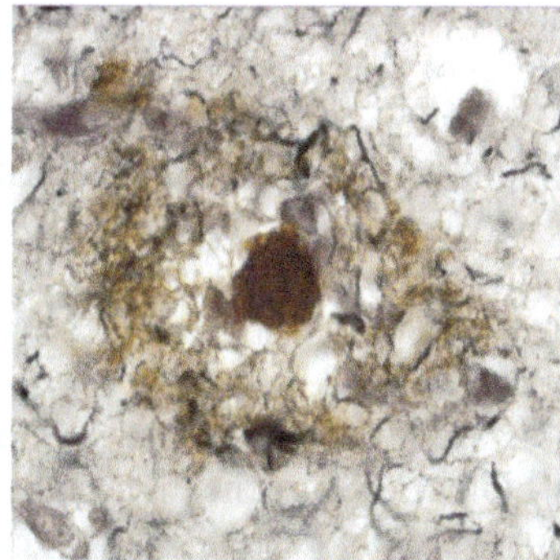

Figure 4.1 Neuritic plaques and neurofibrillary tangles.
(Adapted from G. Deckert *et al.*, in *Nervensystem*, ed. U.-N. Riede, M. Werner, H.-E. Schäfer (Allgemeine und Spezielle Pathologie, vol. 5), Thieme, Stuttgart, 2004, pp. 1039–1114).

serious brain injuries. In-depth, the effect of oxygen deprivation induces a decrease in mitochondrial electron transport chain (ETC) activities and as a result mitochondrial membrane potential loss, reduction in glycolytic ATP and mitochondrial Ca^{2+} release, which all critically contribute to raise neuronal cells stress and subsequently degeneration.[4]

On the other hand, aberrant levels of biological metals in the brain have been found to contribute to AD pathogenesis, by the generation of reactive oxygen species (ROS).[5] Indeed, ROS-induced intracellular calcium influx leads to activation of glutamate receptors and neuronal cell death. Furthermore, metals can contribute to AD by accelerating protein aggregation. For instance, the cumulative levels of zinc in the brain induce Aβ clump formation similar to the plaques of AD.[6]

Abnormal homeostasis of certain biological metals in the brain is a contributing factor in AD. It appears that disruption of the sensitive metal balance in the brain or the accumulation of toxic metals may drive AD pathogenesis. However, the triggering factors and the period of their action increase the complexity and the difficulty of accurately predicting the occurrence of this disease. In this chapter, our aim is to analyse the role of metals in the progression of AD.

4.2 Definition of Alzheimer's Disease and Clinical Presentation

Memory loss and modest changes in behaviour are the standard manifestations of AD in its early stages.[7] Initial memory deficits present as difficulty in forming and retaining memories of recent events.[8] More remote memories remain, at this time, relatively unscathed. With the disease's slow, insidious progression – over the course of some five to 20 years[9] – the patient's episodic, semantic and working memories are ablated, and he/she descends into total dementia. Abnormality of gait and behaviour, difficulties with orientation, major dysfunction of language and reasoning, and incontinence cheat the Alzheimer's patient of their independence and dignity.[9] As such, this disease has considerable emotional and practical implications for the affected and their families. At present, no curative treatment exists for this devastating disease.

The diagnosis of Alzheimer's disease is made on the basis of clinical symptoms supported by mental status examination; the criteria for these (originally outlined in 1984, and since refined) are well established, and diagnoses are generally accurate.[10] However, in the absence of any definitive tests or biomarkers for the disease, a definitive diagnosis can only be confirmed upon post-mortem pathological investigation.[11]

4.3   Genetic Contribution

Six years have passed since the centenary of Alzheimer's neat delineation of his eponymous disease, and still its aetiology remains an incomplete picture. After advanced age, a positive family history is the greatest risk factor for developing AD. Alzheimer's is labelled a "complex" disease on the basis that no single mode of inheritance underpins its heritability: a number of mutations and polymorphisms in a handful of key genes have been identified in the last two decades and their interaction with non-genetic factors contributes to the heterogeneous nature of Alzheimer's disease.[12]

The first genetic mutation implicated in Alzheimer's disease was identified – in 1990 – in the gene encoding the amyloid precursor protein (APP) located on chromosome 21; since then a further 20 pathogenic missense mutations have been discovered in this same gene, all located close to, or within, the domain for the Aβ peptide (Figure 4.2), a key player in the pathology of AD.[12] Mutations in the APP gene are estimated to account for up to 5% of familial AD (FAD). All APP mutations identified so far increase levels of these toxic Aβ species, but in 2012 a new APP mutation was identified which acts in the opposite way, by reducing peptide accumulation and protecting against neurodegeneration.[13] Other genetic mutations causing early-onset familial AD (EOFAD) were identified in presenilin 1 and 2,[14] located on chromosomes 14 and 1, respectively. All these mutations are rare and transmitted in a mendelian fashion[15] and are linked to increases in Aβ levels, particularly Aβ1–42.

Other genetic forms of AD are classified as "sporadic" cases, with less apparent or no familial aggregation and usually of later onset age (>60 years, late-onset AD [LOAD]).[16] One of the first such "candidate genes" assessed for genetic association with AD was APOE (encoding apolipoprotein E [apoE]) on chromosome 19q13. The original discovery that the ε4 allele of a three allele haplotype (composed of ε2, ε3 and ε4 alleles, which show different biochemical properties at the protein level) leads to a dose-dependent four-fold increase in AD risk compared to noncarriers.[16,17] The LOAD gene APOE ε4 allele is thought to mediate its effect by reducing clearance of extracellular Aβ.[17]

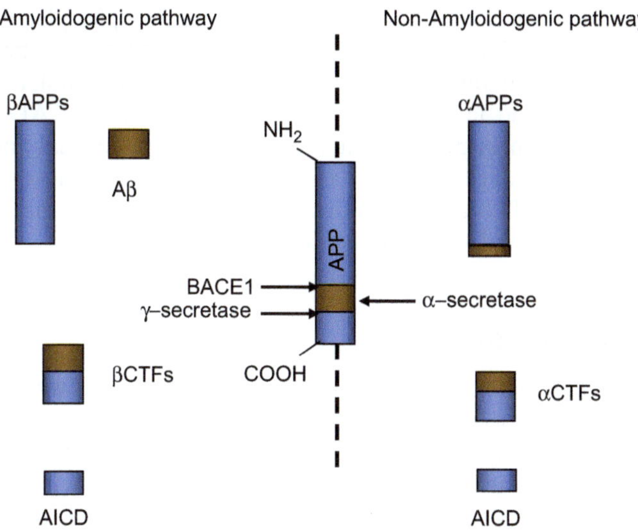

Figure 4.2 Proteolitic processing of the APP.
Adapted from Sastre, *J. Pharmacol. Toxicol. Methods*, 2010, **61**(2), 86–91.

Interestingly, accumulating evidence suggests that LOAD not only results from the combined effects of variation in a number of genes and environmental factors, but also from epigenetic abnormalities such as histone modifications or DNA methylation.[18]

Recent genome-wide screening approaches have revealed several additional AD susceptibility loci and more are likely to be discovered over the coming years;[15] however, they exert only very small risk effects.

In the next section, the significance of Aβ will become clear as we examine the pathology of AD. Thus, identification of these genes and understanding of their effects *in vivo* provide us with clues as to the pathogenesis of AD, and hint at a starting point for efforts in deciphering the underlying mechanisms of the disease.

4.4 Proteins Involved in Alzheimer's Disease

The post-mortem findings in the brain of an AD patient are stereotyped and adherent to well-characterized pathological hallmarks. Levels of the peptide Aβ are increased: deposited extracellularly in characteristic diffuse and neuritic plaques, they are often found surrounded by dystrophic neurites (Figure 4.1). The C-terminal heterogeneity of Aß has special importance for its aggregation. Immunohistochemistry with antibodies that selectively recognize either the valine 40 or the alanine 42 C-terminus has revealed that the first Aß form deposited as diffuse plaques in AD and Down's syndrome brains' ends at residue 42.[9] Evidence has further demonstrated that Aβ also accumulates within neurons, a likely contributor to neuronal dysfunction.[19] The other main intracellular event is the accumulation of hyperphosphorylated tau – a microtubule-associated protein – to form NFTs.[20] Whilst the brain of the patient presents with many other changes, the aberrant accumulation of these two proteins – Aβ and tau – are the two cardinal features of AD.[21] Beyond anomalous deposition of proteins, there is widespread neuronal and synaptic loss, and significant, generalized, atrophy of the cerebral cortices and hippocampus. The AD brain also shows a violent glial disturbance, with local reactive microgliosis and strong activation of astrocytes.[1]

4.5 Neuroinflammation

Acute phase inflammation in response to injury is associated with the healing process; however, persistent inflammation is hugely damaging, and serves to exacerbate underlying pathology in many diseases.[22] The Alzheimer's brain is one such chronically inflamed environment, characterized by an abundance of reactive microglia and astrocytes.[23,24] Focal activation of microglia and astrocytes in the vicinity of Aβ plaques suggests that extracellular Aβ deposition is responsible for triggering the inflammatory reaction.[23]

Microglia release a host of proinflammatory factors, among them pro-inflammatory cytokines, reactive oxygen species (ROS) and nitric oxide (NO), that inflict oxidative and nitrosative injury on the brain. However, recent experimental evidence suggests that there are alternative activation states of microglia in the brain,[25] which may be either beneficial or detrimental depending on the timing and duration of the activation.[26] Astrocytes respond to the long-term inflammation by undergoing transformation from a basal state, in which they are relatively quiescent, to a state of heightened activity, or "activation";[24] furthermore, they proliferate in the inflamed environment of the AD brain and display a hypertrophic phenotype. The astrocytes begin to secrete a variety of proinflammatory mediators under these conditions, undertaking a novel role as immunological cells. In addition to inflammatory molecules, activated astrocytes can release neurotoxic proinflammatory cytokines, ROS and NO.[22]

4.6 Involvement of Metals in AD

4.6.1 Levels of Metals in AD Patients' Brain, CSF and Plasma

The most abundant biochemical metals in eukaryotes are iron, zinc and copper. There is evidence which suggests that copper, iron and zinc are enriched within the amyloid plaques, leaving the brain tissue and cells deficient in these metals.[27,28] This is relevant because maintenance of metal homeostasis is crucial for neuronal functioning.

The blood–brain barrier (BBB) flux is important for aluminium and other metals such as manganese and iron transported into the brain. In fact, evidence suggests that brain aluminium and iron uptake is mediated by a transferrin-receptor dependent mechanism and their brain accumulation is suspected to contribute to metal-induced neurodegeneration.[29] In addition, the post-mortem brain of AD patients shows differences in proteins involved in the storage of metals, such as ferritin. This intracellular iron-storage protein is increased in microglia and senile plaques.[30] Transferrin is found in senile plaques as well, at increased concentration.[31]

Publications on copper levels in the CSF and plasma of AD patients compared to healthy controls have shown a tendency towards an increase with aging[28,32] and this effect is exacerbated in AD.[33,34] However, it has been described that certain metals such as copper may be decreased upon both Aβ and amyloid precursor protein (APP) expression.[35]

Conversely, Zn ions in serum and blood are decreased in AD[36] but increased in CSF.[37] Reports on Fe levels in plasma, serum and CSF are inconsistent.[38]

Other metals of interest in AD, such as cobalt, have been involved in this disease as an important component of vitamin B12, the deficiency of which is associated with an increased risk of AD by affecting homocysteine and folate levels.[39,40]

Chromium and manganese levels are inversely correlated with Aβ42 levels in the CSF of AD patients,[41] while in AD serum, chromium, cobalt, selenium and iron were positively correlated with cognition.[38] Other metals have been linked to AD because of their high toxicity, such as arsenic, lead, mercury and aluminium, most of them with no biological function. There are conflicting results regarding the levels of these metals in AD tissues, blood and serum.[28,42]

4.6.2 Mechanisms of Metal Toxicity in AD

4.6.2.1 Aβ Synthesis and Degradation

Alterations in the distribution and levels of proteins involved in APP processing, such as APP and presenilins, have been also observed upon metal exposure.[43] In fact, exposure to some heavy metals during brain development predetermined the expression, regulation and processing of APP later in life, and potentially influences the course of amyloidogenesis and oxidative damage.[44]

Copper has been reported to regulate APP expression.[45] Elevated copper levels in cells increase APP mRNA and, conversely, decreased copper levels reduce APP transcription.[46] There is evidence that other metals such as arsenic also increase APP levels.[47]

APP is processed through the alternative cleavage carried out by the proteolytic enzymes called secretases which exist in three principle forms: α-, β- and γ-secretase (Figure 4.2). Proteolysis of APP by α-secretase or β-secretase leads to the secretion of soluble α-APPs or β-APPs. Both secretases generate C-terminal fragments of 10 kDa and 12 kDa, respectively, which are inserted in the membrane. These fragments can be cut by γ-secretase to release the peptides P3 and Aβ and a cytoplasmic fragment identified as AICD (APP intracellular domain).[48]

The activity of the secretases may be influenced by the levels of certain metals. For instance, the family of the related metalloproteases of the ADAM (a disintegrin and metalloprotease) family, ADAM-9, ADAM-10 and ADAM-17, which have been identified as the main α-secretases, are regulated by zinc.[49] Iron can also modulate the activation of α-secretases through furin, which is involved in its prodomain cleavage.[50]

β-Secretase (BACE1 for β-site APP cleaving enzyme) was identified as a transmembrane aspartyl protease, which cleaves APP at the N-terminal position of Aβ. BACE1 has been found to interact with copper.[51] In addition, copper upregulates BACE1 expression in PC12 cells.[52] Copper deficiency seems to increase Aβ levels by affecting its synthesis as well as its degradation in lipid rafts.[53] However, increases in intracellular copper by diet or by phar-macological manipulation decreases Aβ and increases sAPPα.[53]

Regarding γ-secretase, it has been published that zinc is important in regu-lating the synthesis of presenilin, the subunit that contains the catalytic site of γ-secretase.[54]

The effects of copper on Aβ degradation have been linked to the stimu-lation of the PI3K pathway, which promotes upregulation of matrix metal-loproteases 2 and 3 (MMP2 and MMP3), which are involved in Aβ clearance.[55] On the other hand, copper was found to downregulate neprilysin activity through modulation of neprilysin protein degradation in N2a cells.[56] Zn^{2+} also regulates the activity of proteins involved in Aβ clearance (like MMPs).[57] The main metalloprotease involved in Aβ degradation in the brain is the insulin-degrading enzyme (IDE), which is a conserved Zn^{2+} metalloendopeptidase. Mass spectrometry and kinetic studies revealed that, among all the metal ions tested, only Cu^{2+}, Cu^+ and Ag^+ have an inhibitory effect on IDE activity. Moreover, the inhibition of copper(II) was reversed by adding zinc(II), whereas the monovalent cations affect the enzyme activity irreversibly.[58]

4.6.2.2   *Effect of Metals on Aβ Toxicity and Aggregation*

Copper, zinc and iron are found in high micromolar concentrations in amyloid plaques.[59] The Aβ sequence includes histidine residues at positions 6, 13 and 14, which enables coordination of transition metal ions.[60] Aβ is able to bind copper, zinc and iron with different affinity,[61,62] which may vary depending on the aggregated state and length of Aβ, and the pH. A greater affinity of copper for Aβ42 over smaller Aβ peptides has been reported,[28] enhancing its precipitation and toxicity.

Metal–Aβ interactions are linked to neurotoxicity by two mechanisms: (i) facilitating Aβ aggregation and (ii) inducing oxidative stress through ROS generation, leading to neuronal cell death and cognitive impairment.[59]

Aggregation of Aβ is fundamental for its toxicity. It is now widely accepted that oligomeric forms are the most toxic form of the peptide. Metals are essential for the formation of soluble aggregates. Particularly, the interaction between Aβ and Cu^{2+} promotes oligomerization and neurotoxicity through the creation of dityrosine covalent bonds between $A\beta$ species as a consequence of copper-mediated conformational change of the $A\beta$ protein.[62,63]

By contrast, zinc seems to precipitate $A\beta$ in a neuroprotective manner.[64] Zn^{2+}-dependent proteins are involved in $A\beta$ clearance (like MMPs) or free radical scavenging (*e.g.* metallothioneins); therefore Zn^{2+} deficiency may be deleterious.[57] However, excessive zinc facilitates $A\beta$ oligomerization and the production of ROS.

Iron was reported to promote toxicity within experimental models of AD through its ability to delay the formation of amyloid aggregates in plaques.[65]

The neurotoxic effects that could be induced by interactions between metals and $A\beta$ may be caused by disruption of membranes, leading to apoptosis and/or aggregate accumulation at the synapse, affecting signalling.[66] In fact, $A\beta$ has been found in mitochondrial membrane and accumulated cytosolic Zn^{2+} increases accumulation of ROS, subsequently reducing mitochondria protection by $A\beta$ mitochondrial aggregation. This destabilizes mitochondria integrity through membrane permeability.[67]

Interestingly, the rat and mouse $A\beta$ sequences have structural changes in some amino acids that mitigate metal ion coordination and lower the aggregation propensity. This could be the reason why these animals do not form cerebral $A\beta$ deposits unless genetically modified to carry the humanized $A\beta$ sequences.[28]

When metal ions are surrounded by $A\beta$ species, ROS such as H_2O_2 and hydroxyl radicals can be generated. The overproduction of ROS can affect and damage lipids, proteins and DNA as well as cause oxidation of $A\beta$ peptides, which could, at the same time, accelerate its aggregation.[59] For instance, enrichment of copper and $A\beta$ in lipid rafts promotes the formation of redox-active $A\beta$:Cu^{2+} complexes which could catalytically oxidize cholesterol, concentrated in lipid rafts, which could then generate H_2O_2 and toxic oxysterols.[27]

Conversely, $A\beta$ monomers can act as antioxidants by detoxifying ROS generated by Cu^{2+} and Fe^{3+} redox activities.[68]

Metal chelators have been used to sequester or redistribute metal ions from metal-bound $A\beta$ species in order to suppress meta-mediated $A\beta$ aggregation *in vitro* and *in vivo*.[59] The use of these chelators aids in the extraction of $A\beta$ peptides from human AD brain samples, by disaggregating oligomers.

4.6.2.3 Metals Involvement in Tau Aggregation

NFTs are the other hallmark of AD, which are aggregates of hyperphosphorylated tau. *In vitro* studies have shown that tau can bind to copper and iron in a similar fashion to their binding to $A\beta$ and contribute to neuronal oxidative stress.[69] Copper binding contributes to the change of conformation, which promotes aggregation of this peptide. In addition, the second repeat region of tau is able to reduce Cu^{2+} and produce H_2O_2.[70]

However, the results on the effect of copper on tau *in vivo* are conflicting, showing increases in tau phosphorylation in triple transgenic 3XTg mice treated with copper[71] or decreases in tau phosphorylation and Aβ aggregation in APP/PS1 mice when the animals are treated with a compound that increases copper availability.[55] *In vitro* it has been shown that synthetic copper ligands can increase copper concentrations and inhibit GSK3, reducing tau phosphorylation.[72]

Zn^{2+} also promotes tau aggregation[73] and NFTs contain high levels of Zn^{2+}. Low levels of zinc facilitate tau fibrillization; however, higher levels have the opposite effect.[73]

4.7 Neurotransmitters Implicated

Disturbances in glutamate and acetycholine, which are major neurotransmitters involved in learning and memory, have been linked with AD.

Glutamate is the major fast excitatory neurotransmitter and is involved in almost all CNS functions, especially in cortical and hippocampal regions. The hippocampus, with its high density of glutamate receptors and in particular NMDA receptors, is known to be extremely important for some forms of learning and memory. Glutamatergic synapses can show pronounced plasticity in terms of the number and strength of individual synapses and are also characterized by their ability to express LTP – a long-lasting strengthening of synaptic transmission.[74,75] This remodelling at the cellular and molecular level is widely accepted to be an underlying synaptic mechanism for learning and memory.[75,76] Chronic, mild activation of NMDA receptors ultimately leads to neurodegeneration – an effect termed chronic "excitotoxicity".[77] In addition, there are evidences supporting a specific effect of Aβ on NMDA receptor-dependent learning.

Acetylcholine, a neurotransmitter essential for processing memory and learning, is decreased in both concentration and function in patients with AD. It is believed that in AD, degeneration of cholinergic basal forebrain neurons leads to cholinergic cortical hypofunction and therefore to cognitive decline and profound dementia.[78–80] This deficit and other presynaptic cholinergic deficits, including loss of cholinergic neurons and decreased acetyl-cholinesterase activity, underscore the cholinergic hypothesis of Alzheimer's disease.[81]

In addition, data obtained from post-mortem AD brains and cerebrospinal fluid from patients indicates deficits in the noradrenergic and serotonergic systems as well.[82] Loss of noradrenergic innervation is very pronounced in the locus ceruleus (LC) at early stages of the disease, probably contributing to the emergence of cognitive impairments.[83] The neurons from the LC, located on the lateral aspect of the fourth ventricle, constitute the primary source of norepinephrine (NE) in the central nervous system.[84,85] The degree of cell loss in the LC appears to be more extensive than in any other brain region in AD, which leads to the speculation that the LC changes could be a primary event by itself early in the pathogenesis of AD.[86] Moreover, the decreased neuronal number in the LC significantly correlated with the number of Aβ plaques, neurofibrillary tangles and severity of the dementia in AD.[87]

4.8 Therapeutic Strategies

4.8.1 Current Therapies

The currently available drugs for symptomatic treatment of AD (*i.e.* memantine and acetylcholinesterase inhibitors) only temporarily slow down the natural history of the disease process and are unable to prevent or reverse the disease.[88,89]

4.8.1.1 Acetylcholinesterase Inhibitors

Cholinesterases are a group of serine hydrolases that split the neurotransmitter acetylcholine (ACh) and terminate its action. Of the two types, butyryl-cholinesterase and acetylcholinesterase (AChE), AChE plays the key role in ending cholinergic neurotransmission. Cholinesterase inhibitors are substances, either natural or man-made, that interfere with the breakdown of ACh and prolong its action.[90] The cholinesterase inhibitors (ChEIs) approved to be used to ameliorate AD symptoms are donepezil, galantamine and rivastigmine (Figure 4.3).[88–90]

4.8.1.2 Memantine

Strong support for the clinical relevance of such interactions between Aβ, glutamate and NMDA receptors in AD is provided by the NMDA receptor antagonist memantine. This substance is the only NMDA receptor antagonist used clinically in the treatment of AD.[77] Memantine's modulation of NMDA receptors has been reported to prevent the neuronal necrosis induced by glutamatergic calcium neurotoxicity.[91]

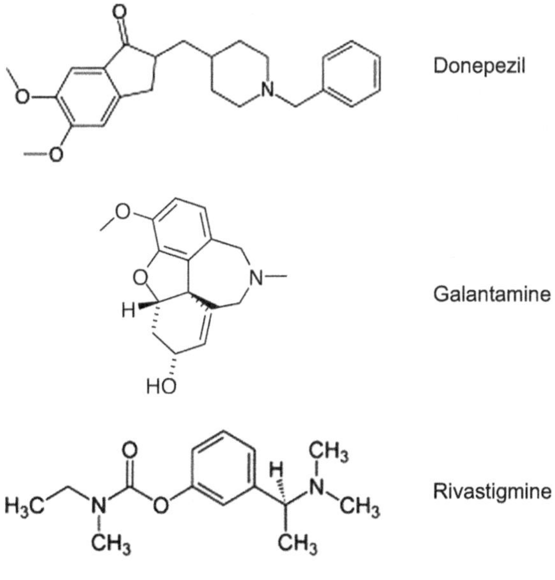

Figure 4.3 Chemical structure of cholinesterase inhibitors.

4.8.2 Therapeutic Strategies Based on Metals

Based on the properties of metals to increase Aβ aggregation and toxicity, one approach for AD therapeutics could be inhibiting the interaction between Aβ and transition metals by using agents which facilitate redistribution of these metals into their appropriate physiological compartments. Chelators have been used to treat other diseases, such as haemochromatosis (Fe) and Wilson's disease (Cu). However, these chelators are not suited to treat AD, because it is controversial whether they cross effectively the BBB and have metal binding affinities way in excess of those needed to chaperone metals between compartments. Indeed, in AD brains it is the changed location of metal ions and not the absolute levels that are problematic. Therefore, chelation and thereby dramatic reduction of metal concentrations centrally as well as systemically would be dangerous due to the multitude of physiological, metalloprotein interactions in the nervous system and elsewhere.[60]

Thus other approaches have been developed.

4.8.2.1 Metal Protein Attenuating Compounds

The prototype metal protein attenuating compound (MPAC) was clioquinol (a retired anti-amoebic), used liberally across the world until the 1970s. It is lipophilic and therefore is able to cross the BBB and has moderate affinity for metal ions so as to affect central concentrations with no peripheral effects (Figure 4.4). Studies in transgenic AD mice have shown reductions in Aβ load, without a systematic decrease in total metal levels.[92] In clinical trials it was shown to have some cognitive benefits over placebo in patients with moderate severity Alzheimer's dementia as well as causing a decrease in plasma $A\beta_{42}$.[93] The effects of clioquinol were also thought to be mediated through increases in Aβ degradation by increasing the expression of MMP2 and MMP3.[94] The development of new MPACs with improved pharmacokinetic and pharmacodynamics properties over clioquinol have led to the phase 2 testing of

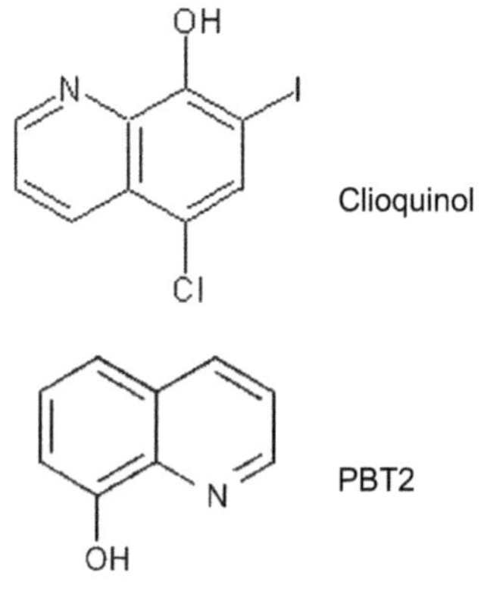

Figure 4.4 Chemical structure of clioquinol and PBT2.

PBT2 (Figure 4.4). In animal models of AD, PBT2 led to decreased Aβ load, tau phosphorylation and improved cognitive performance of transgenic mice on the Morris water maze test.[95] A phase 2 clinical trial also demonstrated a good safety and tolerability profile, significant reductions in CSF Aβ$_{42}$ and improved cognitive performance (executive function) in patients treated with a high dose (250 mg) of PBT2 compared with a low dose (100 mg) and placebo.[96] It seems that the protective effects of these drugs are not only due to the inhibition of Aβ/metal interactions but to induce protective signalling cascades.[94] The impact of PBT2 on cerebral amyloid as indexed by PE-PIB is currently being investigated in Australia in the 12-month duration IMAGINE study and for the treatment of Huntington's disease in the 6-month REACH2HD study. Both studies are due to report by mid-2014.

4.8.2.2 *M{Bis(thiosemicarbazone)} Complexes*

These drugs have a wide range of pharmacological effects. However, they allow delivery of metals in order to increase the intracellular availability of metals such as copper.[60] Complexes of bis(thiosemicarbazone) (btsc) with copper have been proven to be effective in improving cognitive performance in AD mouse models, decreasing the amount of Aβ oligomers and phosphorylated tau.[55]

4.8.2.3 *PtLCl$_2$ Complexes*

A strategy using metal compounds that target the site of binding for Aβ has been developed. A range of platinum-based inhibitors of the Aβ metal binding site, such as PtLCl$_2$ complexes (L = 1,10-phenanthroline derivatives) were examined and shown to bind to Aβ, inhibit neurotoxicity and rescue Aβ-induced synaptotoxicity in mouse hippocampal slices.[97]

4.8.2.4 *Restoring Metal Levels through Dietary Supplementation*

Recent studies confirm that restoring Zn^{2+} levels with dietary supplementation in pre-symptomatic stages of the disease could be beneficial by increasing BDNF levels.[98] However, existing data do not support evidence that depletion or supplementation of copper or iron is beneficial for AD. Prospective studies have shown that a diet with saturated fats and copper or iron can accelerate cognitive decline.[99]

4.9 Research Perspectives in Alzheimer's Disease Linked to Metals and Complexity

While there is documented evidence for a metal link with AD pathogenesis, epidemiologic association is inconclusive; this is mainly due to the complexity of the pathways and the huge number of factors that metals could interact with or regulate. To date, most of the research work has been focused on the interaction between Aβ and metals, although there are many more additional mechanisms in which metals affect neurodegeneration (Figure 4.5). Several critical enzymatic processes associated fundamentally with AD pathology are influenced substantially by metals; however, the molecular basis of this regulation is not yet understood.

Furthermore, metals do not guarantee the development of AD, nor does their absence prevent it. As a new layer of complexity, epigenetics has been shown also to be involved in the modification of AD gene expression together with aging processes and/or environmental factors. Heavy metals and dietary folate intake perturb AD genes by epigenetic means, leading to altered gene expression and late-onset AD.[100] It is likely that epigenetic alterations on DNA, chromatin structure and/or micRNA levels accumulate during AD initiation and development, contributing at least in part to the aetiology of the disease.

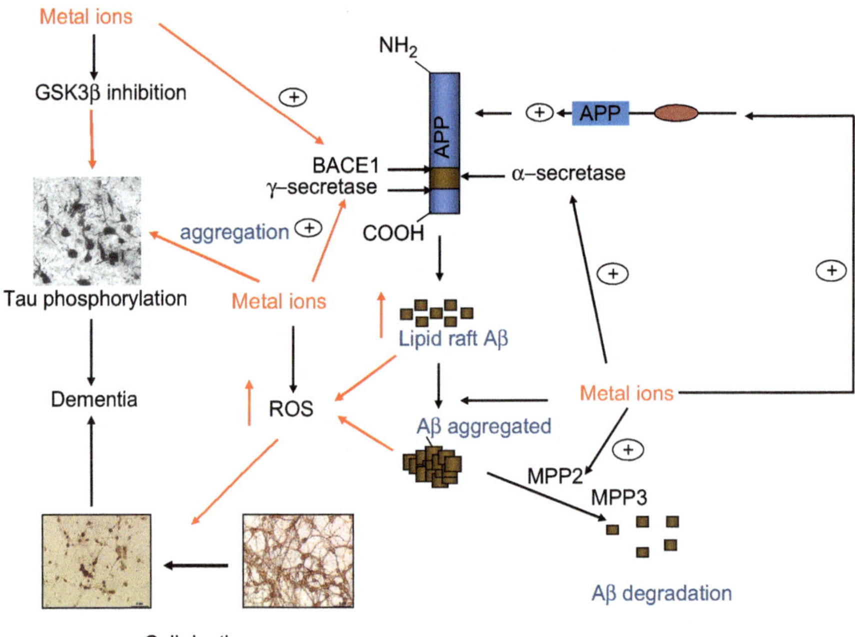

Figure 4.5   Schematic representation of the interactions between metals and AD pathways. Metals affect APP processing by their effects on the expression and distribution of the secretases and APP, therefore influencing Aβ generation. Moreover, metals have been involved in the aggregation of Aβ. On the other hand, certain metals such as Zn can activate Aβ degradation. The increase of Aβ levels and aggregation by metals potentiates the formation of ROS, which are also overproduced by other mechanisms. This results in an increase in neuronal death. Certain metals can also affect tau aggregation by inhibiting GSK3 or by binding directly to tau.

The understanding of this complex disease requires more comprehensive and long-term bio-monitoring which can bridge metals to specific gene risk factors and reveal genetic and epigenetic connections with the disease.

References

1. D. J. Selkoe, *Nature*, 2003, **42**, 900.
2. D. J. Selkoe, *Science*, 2002, **298**, 789.
3. G. Eskici and P. H. Axelsen, *Biochemistry*, 2012, **51**, 6289.
4. E. A. Belyaeva, T. V. Sokolova, L. V. Emelyanova and I. O. Zakharova, *Sci. World J.*, 2012, **2012**, 136063.
5. B. Uttara, A. V. Singh, P. Zamboni and R. T. Mahajan, *Curr. Neuropharmacol.*, 2009, **7**, 65.
6. T. J. A. Craddock, J. A. Tuszynski, D. Chopra, N. Casey, L. E. Goldstein, S. R. Hameroff and R. E. Tanzi, *PloS One*, 2012, **7**, e33552.
7. B. Selman, *Nature*, 2008, **451**, 639.
8. A. E. Budson and B. H. Price, *Pract. Neurol.*, 2007, **7**, 42.
9. D. J. Selkoe and M. B. Podlisny, *Annu. Rev. Genomics Hum. Genet.*, 2002, **3**, 67.
10. G. M. McKhann, D. S. Knopman, H. Chertkow, B. T. Hyman, C. R. Jack, Jr., C. H. Kawas, W. E. Klunk, W. J. Koroshetz, J. J. Manly, R. Mayeux, R. C. Mohs, J. C. Morris, M. N. Rossor, P. Scheltens, M. C. Carrillo, B. Thies, S. Weintraub and C. H. Phelps, *Alzheimer's Dementia*, 2011, **7**, 263.
11. R. Mayeux, *Annu. Rev. Neurosci.*, 2003, **26**, 81.
12. R. E. Tanzi and L. Bertram, *Neuron*, 2001, **32**, 181.
13. T. Jonsson, J. K. Atwal, S. Steinberg, J. Snaedal, P. V. Jonsson, S. Bjornsson, H. Stefansson, P. Sulem, D. Gudbjartsson, J. Maloney, K. Hoyte, A. Gustafson, Y. Liu, Y. Lu, T. Bhangale, R. R. Graham, J. Huttenlocher, G. Bjornsdottir, O. A. Andreassen, E. G. Jönsson, A. Palotie, T. W. Behrens, O. T. Magnusson, A. Kong, U. Thorsteinsdottir, R. J. Watts and K. Stefansson, *Nature*, 2012, **488**, 96.
14. R. E. Tanzi, *J. Clin. Invest.*, 1999, **104**, 1175.
15. L. Bertram and R. E. Tanzi, *Prog. Mol. Biol. Transl. Sci.*, 2012, **107**, 79.
16. L. Bertram, C. M. Lill and R. E. Tanzi, *Neuron*, 2010, **68**, 270.
17. W. J. Strittmatter, K. H. Weisgraber, D. Y. Huang, L. M. Dong, G. S. Salvesen, M. Pericak-Vance, D. Schmechel, A. M. Saunders, D. Goldgaber and A. D. Roses, *Proc. Natl. Acad. Sci. U. S. A.*, 1993, **90**, 8098.
18. S. W. Bihaqi, A. Schumacher, B. Maloney, D. K. Lahiri and N. H. Zawia, *Curr. Alzheimer Res.*, 2012, **9**, 574.
19. G. K. Gouras, J. Tsai, J. Naslund, B. Vincent, M. Edgar, F. Checler, J. P. Greenfield, V. Haroutunian, J. D. Buxbaum, H. Xu, P. Greengard and N. R. Relkin, *Am. J. Pathol.*, 2000, **156**, 20.
20. V. M. Lee, M. Goedert and J. Q. Trojanowski, *Annu. Rev. Neurosci.*, 2001, **24**, 1121.

21. D. P. Perl, *Mt. Sinai J. Med.*, 2010, **77**, 32.
22. M. Sastre, J. C. Richardson, S. M. Gentleman and D. J. Brooks, *Curr. Alzheimer Res.*, 2011, **8**, 132.
23. M. Sastre, T. Klockgether and M. Heneka, *Int. J. Dev. Neurosci.*, 2006, **24**, 167.
24. S. Fuller, M. Steele and G. Münch, *Mutat. Res.*, 2010, **690**, 40.
25. S. Jimenez, D. Baglietto-Vargas, C. Caballero, I. Moreno-Gonzalez, M. Torres, R. Sanchez-Varo, D. Ruano, M. Vizuete, A. Gutierrez and J. Vitorica, *J. Neurosci.*, 2008, **28**, 11650.
26. E. Solito and M. Sastre, *Front. Pharmacol.*, 2012, **3**, 14.
27. Y. H. Hung, A. I. Bush and R. A. Cherny, *J. Biol. Inorg. Chem.*, 2010, **15**, 61.
28. B. R. Roberts, T. M. Ryan, A. I. Bush, C. L. Masters and J. A. Duce, *J. Neurochem.*, 2012, **120**(suppl. 1), 149.
29. M. Kawahara, *J. Alzheimer's Dis.*, 2006, **10**, 223.
30. K. Jellinger, W. Paulus, I. Grundke-Iqbal, P. Riederer and M. B. Youdim, *J. Neural Transm. Park. Dis. Dement. Sect.*, 1990, **2**, 327.
31. J. R. Connor, S. L. Menzies, S. M. St. Martin and E. J. Mufson, *J. Neurosci. Res.*, 1992, **31**, 75.
32. C. Ekmekcioglu, *Nahrung*, 2001, **45**, 309.
33. R. Squitti, D. Lupoi, P. Pasqualetti, G. Dal Forno, F. Vernieri, P. Chiovenda, L. Rossi, M. Cortesi, E. Cassetta and P. M. Rossini, *Neurology*, 2002, **59**, 1153.
34. S. Bucossi, M. Ventriglia, V. Panetta, C. Salustri, P. Pasqualetti, S. Mariani, M. Siotto, P. M. Rossini and R. Squitti, *J. Alzheimer's Dis.*, 2011, **24**, 175.
35. C. J. Maynard, A. I. Bush, C. L. Masters, R. Cappai and Q. X. Li, *Int. J. Exp. Pathol.*, 2005, **86**, 147.
36. G. J. Brewer, S. H. Kanzer, E. A. Zimmerman, E. S. Molho, D. F. Celmins, S. M. Heckman and R. Dick, *Am. J. Alzheimer's Dis. Other Dement.*, 2010, **25**, 572.
37. D. Religa, D. Strozyk, R. A. Cherny, I. Volitakis, V. Haroutunian, B. Winblad, J. Naslund and A. I. Bush, *Neurology*, 2006, **67**, 69.
38. C. Smorgon, E. Mari, A. R. Atti, E. Dalla Nora, P. F. Zamboni, F. Calzoni, A. Passaro and R. Fellin, *Arch. Gerontol. Geriatr. Suppl.*, 2004, **9**, 393.
39. S. Seshadri, A. Beiser, J. Selhub, P. F. Jacques, I. H. Rosenberg, R. B. D'Agostino, P. W. Wilson and P. A. Wolf, *New Engl. J. Med.*, 2002, **346**, 476.
40. C. I. Prodan, L. D. Cowan, J. A. Stoner and E. D. Ross, *J. Neurol. Sci.*, 2009, **284**, 144.
41. D. Strozyk, L. J. Launer, P. A. Adlard, R. A. Cherny, A. Tsatsanis, I. Volitakis, K. Blennow, H. Petrovitch, L. R. White and A. I. Bush, *Neurobiol. Aging*, 2009, **30**, 1069.
42. L. Tomljenovic, *J. Alzheimer's Dis.*, 2011, **23**, 567.
43. M. Nizzari, S. Thellung, A. Corsaro, V. Villa, A. Pagano, C. Porcile, C. Russo and T. Florio, *J. Toxicol.*, 2012, **2012**, 187297.

44. J. Wu, M. R. Basha, B. Brock, D. P. Cox, F. Cardozo-Pelaez, C. A. McPherson, J. Harry, D. C. Rice, B. Maloney, D. Chen, D. K. Lahiri and N. H. Zawia, *J. Neurosci.*, 2008, **28**, 3.
45. S. A. Bellingham, D. K. Lahiri, B. Maloney, S. La Fontaine, G. Multhaup and J. Camakaris, *J. Biol. Chem.*, 2004, **279**, 20378.
46. W. Zheng, N. Xin, Z. H. Chi, B. L. Zhao, J. Zhang, J. Y. Li and Z. Y. Wang, *FASEB J.*, 2009, **23**, 4207.
47. S. Zarazúa, S. Bürger, J. M. Delgado, M. E. Jiménez-Capdeville and R. Schliebs, *Int. J. Dev. Neurosci.*, 2011, **29**, 389.
48. M. Sastre, J. Walter and S. M. Gentleman, *J. Neuroinflammation*, 2008, **5**, 25.
49. D. R. Edwards, M. M. Handsley and C. J. Pennington, *Mol. Aspects Med.*, 2008, **29**, 258.
50. L. Silvestri and C. Camaschella, *J. Cell. Mol. Med.*, 2008, **12**, 1548.
51. B. Angeletti, K. J. Waldron, K. B. Freeman, H. Bawagan, I. Hussain, C. C. Miller, K. F. Lau, M. E. Tennant, C. Dennison, N. J. Robinson and C. Dingwall, *J. Biol. Chem.*, 2005, **280**, 17930.
52. R. Lin, X. Chen, W. Li, Y. Han, P. Liu and R. Pi, *Neurosci. Lett.*, 2008, **440**, 344.
53. M. A. Cater, K. T. McInnes, Q. X. Li, I. Volitakis, S. La Fontaine, J. F. Mercer and A. I. Bush, *Biochem. J.*, 2008, **412**, 141.
54. I. H. Park, M. W. Jung, H. Mori and I. Mook-Jung, *Biochem. Biophys. Res. Commun.*, 2001, **285**, 680.
55. P. J. Crouch, L. W. Hung, P. A. Adlard, M. Cortes, V. Lal, G. Filiz, K. A. Perez, M. Nurjono, A. Caragounis, T. Du, K. Laughton, I. Volitakis, A. I. Bush, Q. X. Li, C. L. Masters, R. Cappai, R. A. Cherny, P. S. Donnelly, A. R. White and K. J. Barnham, *Proc. Natl. Acad. Sci. U. S. A.*, 2009, **106**, 381.
56. M. Li, M. Sun, Y. Liu, J. Yu, H. Yang, D. Fan and D. Chui, *J. Alzheimer's Dis.*, 2010, **19**, 161.
57. C. Corona, A. Pensalfini, V. Frazzini and S. L. Sensi, *Cell Death Dis.*, 2011, **2**, 176.
58. G. Grasso, A. Pietropaolo, G. Spoto, G. Pappalardo, G. R. Tundo, C. Ciaccio, M. Coletta and E. Rizzarelli, *Chemistry*, 2011, **17**, 2752.
59. A. S. Pithadia and M. H. Lim, *Curr. Opin. Chem. Biol.*, 2012, **16**, 67.
60. V. B. Kenche and K. J. Barnham, *Br. J. Pharmacol.*, 2011, **163**, 211.
61. A. I. Bush, W. H. Pettingell, G. Multhaup, M. d'Paradis, J. P. Vonsattel, J. F. Gusella, K. Beyreuther, C. L. Masters and R. E. Tanzi, *Science*, 1994, **265**, 1464.
62. C. S. Atwood, R. C. Scarpa, X. Huang, R. D. Moir, W. D. Jones, D. P. Fairlie, R. E. Tanzi and A. I. Bush, *J. Neurochem.*, 2000, **75**, 1219.
63. C. S. Atwood, R. D. Moir, X. Huang, R. C. Scarpa, N. M. Bacarra, D. M. Romano, M. A. Hartshorn, R. E. Tanzi and A. I. Bush, *J. Biol. Chem.*, 1998, **273**, 12817.
64. M. P. Cuajungco, L. E. Goldstein, A. Nunomura, M. A. Smith, J. T. Lim, C. S. Atwood, X. Huang, Y. W. Farrag, G. Perry and A. I. Bush, *J. Biol. Chem.*, 2000, **275**, 19439.

65. B. Liu, A. Moloney, S. Meehan, K. Morris, S. E. Thomas, L. C. Serpell, R. Hider, S. J. Marciniak, D. A. Lomas and D. C. Crowther, *J. Biol. Chem.*, 2011, **286**, 4248.

66. A. Rauk, *Chem. Soc. Rev.*, 2009, **38**, 2698.

67. K. P. Kepp, *Chem. Rev.*, 2012, **112**, 5193.

68. K. Zou, J. S. Gong, K. Yanagisawa and M. Michikawa, *J. Neurosci.*, 2002, **22**, 4833.

69. L. M. Sayre, G. Perry, P. L. Harris, Y. Liu, K. A. Schubert and M. A. Smith, *J. Neurochem.*, 2000, **74**, 270.

70. X. Y. Su, W. H. Wu, Z. P. Huang, J. Hu, P. Lei, C. H. Yu, Y. F. Zhao and Y. M. Li, *Biochem. Biophys. Res. Commun.*, 2007, **358**, 661.

71. M. Kitazawa, D. Cheng and F. M. Laferla, *J. Neurochem.*, 2009, **108**, 1550.

72. J. L. Hickey, P. J. Crouch, S. Mey, A. Caragounis, J. M. White, A. R. White and P. S. Donnelly, *Dalton Trans.*, 2011, **40**, 1338.

73. Z. Y. Mo, Y. Z. Zhu, H. L. Zhu, J. B. Fan, J. Chen and Y. Liang, *J. Biol. Chem.*, 2009, **284**, 34648.

74. C. W. Cotman, D. T. Monaghan and A. H. Ganong, *Annu. Rev. Neurosci.*, 1988, **11**, 61.

75. G. L. Collingridge and W. Singer, *Trends Pharmacol. Sci.*, 1990, **11**, 290.

76. D. A. Butterfield and C. B. Pocernich, *CNS Drugs*, 2003, **17**, 641.

77. W. Danysz and C. G. Parsons, *Br. J. Pharmacol.*, 2012, **167**, 324.

78. P. J. Whitehouse, D. L. Price, A. W. Clark, J. T. Coyle and M. R. DeLong, *Ann. Neurol.*, 1981, **10**, 122.

79. P. J. Whitehouse, D. L. Price, R. G. Struble, A. W. Clark, J. T. Coyle and M. R. Delon, *Science*, 1982, **215**, 1237.

80. A. M. Palmer, *Neurodegeneration*, 1996, **5**, 381.

81. P. T. Francis, *CNS Spectr.*, 2005, **10**, 6.

82. A. M. Palmer and S. T. DeKosky, *J. Neural Transm. Gen. Sect.*, 1993, **91**, 135.

83. M. R. Marien, F. C. Colpaert and A. C. Rosenquist, *Brain Res. Brain Res. Rev.*, 2004, **45**, 38.

84. G. Aston-Jones, J. Rajkowski, P. Kubiak, R. J. Valentino and M. T. Shipley, *Prog. Brain Res.*, 1996, **107**, 379.

85. G. Aston-Jones, J. Rajkowski and J. Cohen, *Prog. Brain Res.*, 2000, **126**, 165.

86. D. C. German, K. F. Manaye, C. L. White, 3rd, D. J. Woodward, D. D. McIntire, W. K. Smith, R. N. Kalaria and D. M. Mann, *Ann. Neurol.*, 1992, **32**, 667.

87. W. Bondareff, C. Q. Mountjoy, M. Roth, M. N. Rossor, L. L. Iversen, G. P. Reynolds and D. L. Hauser, *Alzheimer Dis. Assoc. Disord.*, 1987, **1**, 256.

88. I. Melkinova, *Nat. Rev.*, 2007, **6**, 341.

89. R. J. van Marum, *Fundam. Clin. Pharmacol.*, 2008, **22**, 265.

90. M. Pohanka, *Expert Opin. Ther. Pat.*, 2012, **22**, 871.

91. C. Annweiler and O. Beauchet, *Drugs Aging*, 2012, **29**, 81.

92. R. A. Cherny, C. S. Atwood, M. E. Xilinas, D. N. Gray, W. D. Jones, C. A. McLean, K. J. Barnham, I. Volitakis, F. W. Fraser, Y. Kim, X. Huang, L. E. Goldstein, R. D. Moir, J. T. Lim, K. Beyreuther, H. Zheng, R. E. Tanzi, C. L. Masters and A. I. Bush, *Neuron*, 2001, **30**, 665.

93. C. W. Ritchie, A. I. Bush, A. Mackinnon, S. Macfarlane, M. Mastwyk, L. MacGregor, L. Kiers, R. Cherny, Q. X. Li, A. Tammer, D. Carrington, C. Mavros, I. Volitakis, M. Xilinas, D. Ames, S. Davis, K. Beyreuther, R. E. Tanzi and C. L. Masters, *Arch. Neurol.*, 2003, **60**, 1685.

94. A. R. White, T. Du, K. M. Laughton, I. Volitakis, R. A. Sharples, M. E. Xilinas, D. E. Hoke, R. M. Holsinger, G. Evin, R. A. Cherny, A. F. Hill, K. J. Barnham, Q. X. Li, A. I. Bush and C. L. Masters, *J. Biol. Chem.*, 2006, **281**, 17670.

95. P. A. Adlard, R. A. Cherny, D. I. Finkelstein, E. Gautier, E. Robb, M. Cortes, I. Volitakis, X. Liu, J. P. Smith, K. Perez, K. Laughton, Q. X. Li, S. A. Charman, J. A. Nicolazzo, S. Wilkins, K. Deleva, T. Lynch, G. Kok, C. W. Ritchie, R. E. Tanzi, R. Cappai, C. L. Masters, K. J. Barnham and A. I. Bush, *Neuron*, 2008, **59**, 43.

96. L. Lannfelt, K. Blennow, H. Zetterberg, S. Batsman, D. Ames, J. Harrison, C. L. Masters, S. Targum, A. I. Bush, R. Murdoch, J. Wilson, C. W. Ritchie and PBT2-201-EURO study group, *Lancet Neurol.*, 2008, **7**, 779.

97. K. J. Barnham, V. B. Kenche, G. D. Ciccotosto, D. P. Smith, D. J. Tew, X. Liu, K. Perez, G. A. Cranston, T. J. Johanssen, I. Volitakis, A. I. Bush, C. L. Masters, A. R. White, J. P. Smith, R. A. Cherny and R. Cappai, *Proc. Natl. Acad. Sci. U. S. A.*, 2008, **105**, 6813.

98. C. Corona, F. Masciopinto, E. Silvestri, A. D. Viscovo, R. Lattanzio, R. L. Sorda, D. Ciavardelli, F. Goglia, M. Piantelli, L. M. Canzoniero and S. L. Sensi, *Cell Death Dis.*, 2010, **1**, 91.

99. M. Loef and H. Walach, *Br. J. Nutr.*, 2012, **107**, 7.

100. J. B. Kwok, *Epigenomics*, 2010, **2**, 671.

CHAPTER 5

Friedreich's Ataxia

ANNALISA PASTORE

MRC National Institute for Medical Research
Email: apastor@nimr.mrc.ac.uk

5.1   Introduction

Germany was the cradle of modern neurology. The German school of medicine
in the 19th–20th century was the reference point for the diagnosis of several
diseases, among which were several neurodegenerative and neurological
diseases. The most commonly known example is the work of Alois Alzheimer
(1864–1915), who at the turn of the 20th century diagnosed for the first time the
disease which now carries his name.[1] Less famous but no less pioneering was
the work of Nicholaus Friedreich, a professor of medicine in Heidelberg who
described for the first time in 1863 a rare neurodegenerative disease now known
as Friedreich's ataxia (FRDA)[2] (MIM 229300). He described the essential
clinical and pathological features of this disease as characterized by "degen-
erative atrophy of the posterior columns of the spinal cord". After its first
description, more than a century was needed to achieve substantial progress in
understanding the molecular bases of the disease. In this chapter, we analyse
our current understanding of the genetic and biochemical causes of FRDA and
discuss possible therapeutic approaches.

5.2   Definition of the Disease and Clinical Presentation

Diagnostic criteria were established in the late 1970s, mainly by G. Geoffroy
and colleagues[3] and A. E. Harding.[4,5] The latter identified, as a necessary
condition for correct diagnosis of FRDA, autosomal recessive inheritance and

RSC Metallobiology Series No. 1
Mechanisms and Metal Involvement in Neurodegenerative Diseases
Edited by Roberta Ward, David Dexter and Robert Crichton

Published by the Royal Society of Chemistry, www.rsc.org

disease onset before age 25. We now know, however, that while these criteria identify the most typical cases of FRDA, the disease may have large clinical variability in age of onset, rate of progression and severity of disease.

Most neurological symptoms are a consequence of degeneration of the large sensory neurons of dorsal root ganglia and spinocerebellar tracts.[6] This state leads to progressive ataxia, sensory loss and muscle weakness, symptoms which are often accompanied by scoliosis, foot deformity and optic atrophy. The most important non-neurological symptoms are hypertrophic cardiomyopathy and increased incidence of diabetes mellitus, which are often the causes of death.

FRDA is the most common of the hereditary ataxias and yet it is classified as a rare disease, accounting for *ca.* one case in 50 000 individuals in Caucasian populations.[7,8] However, being recessive, it has an estimated carrier prevalence of one in 110 individuals. The first symptoms usually develop in childhood or puberty, but the age of onset can vary from infancy to adulthood. Life expectancy averages between 40 and 50 years.

5.3   Genetic Contribution

The genetic locus of FRDA was first mapped to chromosome 9q13 by Chamberlain and colleagues.[9,10] Identification of the specific responsible gene was achieved thanks to an international collaboration that linked FRDA to a previously uncharacterized transcript originally named *X25*.[11] The *FRDA* gene comprises five exons, which allow formation of up to seven transcripts (Figure 5.1).[11,12] The most common mutation in FRDA is an expanded GAA

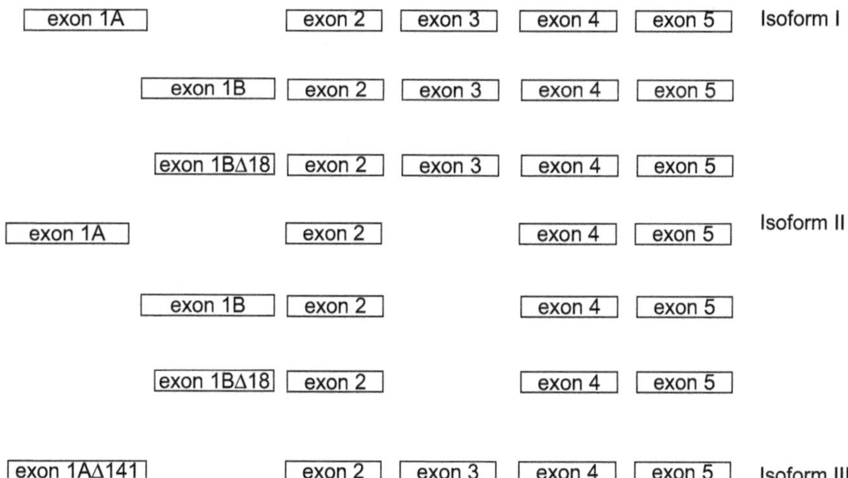

Figure 5.1   Schematic diagram of the different transcripts of human frataxin as identified by PCR. Exon1A is the first canonical exon; exon1AΔ141 contains exon1A missing the last 141 nucleotides; exon1B is an alternative to exon 1; exon1BΔ18 misses 18 nucleotides near the 5'-end of exon1B. (Adapted from Xia *et al.*[12]).

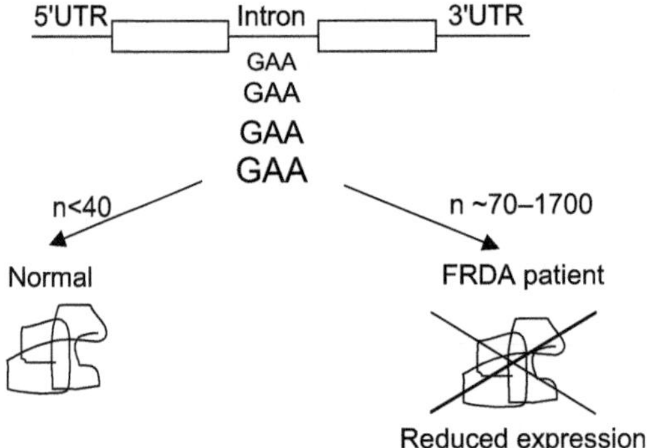

Figure 5.2 Genetic causes of FRDA. The disease is usually caused by anomalous expansion of a GAA triplet in the first intron of the gene. The expansion leads to partial silencing of the protein frataxin. The pathological threshold is approximately above 30–40 repeats.

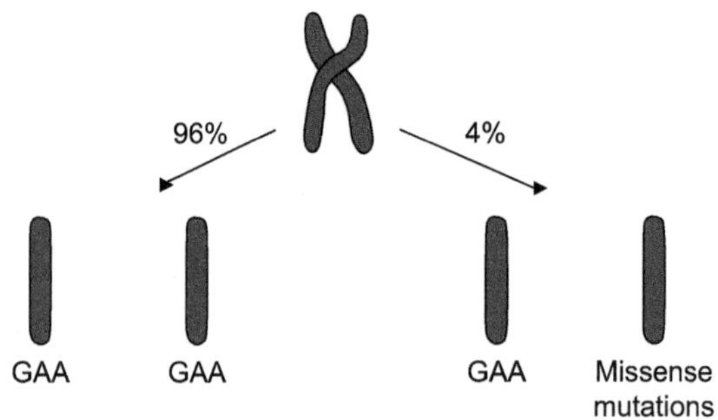

Figure 5.3 Genetic causes of FRDA. Most of the FRDA patients (96%) have two expanded alleles, while a small minority (4%) have the expansion on one allele and missense mutations on the other.[11] There are no cases of homozygous for missense mutations.

trinucleotide repeat in intron 1 of the *FRDA* gene (Figure 5.2).[11,13,14] Healthy individuals are characterized by 7–36 GAA repeats, while FRDA patients carry 70–1700 GAA repeats.[11,13–17] The severity of FRDA and the age of onset correlate with the number of repeats.[13] FRDA is the first example of an autosomal recessive disease caused by anomalous expansion of a dynamic GAA trinucleotide repeat. Several other diseases are known to be associated with triplet expansion but they are in most cases dominant.[18,19] It is interesting to compare the extent of the non-coding expansion in FRDA with the much more modest exonic repeat expansions observed in some of these dominant diseases (*e.g.* spinocerebellar ataxias and Huntington's chorea), where pathology manifests as soon as expansion exceeds a threshold of *ca.* 37–50 repeats.[15]

Most FRDA patients (96%) carry two expanded alleles (Figure 5.3).[11,20] A smaller number of cases (4% of the patients) carry expansion on only one allele, whereas the other allele contains missense mutations.[21,22] No FRDA patient has so far been found to be homozygous for point mutation, suggesting that the gene product must be essential for survival. This hypothesis is fully confirmed by the observation that complete knockout of the *FRDA* gene in mice models is embryonically lethal.[20]

The gene product of *FRDA* was called frataxin as a fusion name between FRDA and ataxia (by the time frataxin was identified, proteins responsible for dominant spinocerebellar ataxias had already been named ataxins).[11] The non-coding GAA expansion leads to reduced levels of frataxin, making this disease a clear example of a loss-of-function mechanism, compared with the vast number of neurodegenerative (and non-neurodegenerative) diseases that are caused by gain of a toxic function due to protein aggregation and misfolding.[23–25] Two non-mutually exclusive models seem to be widely accepted to explain how GAA expansion may induce partial transcriptional inhibition of frataxin. They are based on formation of non-B DNA conformations and/or on a heterochromatin-mediated mechanism. The potential crosstalk between the two mechanisms is currently still unclear and is the subject of further studies.

Solid experimental evidence *in vitro* and *in vivo* indicates that DNA containing GAA repeats adopts intramolecular non-B triple helical structures that lock the DNA molecule into a dumbbell-shaped structure that directly interferes with transcriptional elongation.[26–28] This is the direct consequence of the complementarity between strands with only purines (R) and strands with only pyrimidines (Y), which allows the intercalation of a third strand. Long (above *ca.* 59 and up to *ca.* 300 repeats) RRY triplexes can further adopt higher-order intramolecular conformations called "sticky DNA".[29–33] These very stable dumbbell-shaped structures block transcription, replication, repair and recombination and prevent nucleosome assembly.

The second model is based on a heterochromatin-mediated gene silencing model. Although the mechanism is still unclear, the phenomenon has been compared to position effect variegated gene silencing, a well-known way to achieve silencing of genes located within or near regions of hetero-chromatin.[34] Independent studies have in fact indicated that a hetero-chromatin structure is present in the region around the GAA expansion and could inhibit tran-scription of the *FRDA* gene. Hetero-chromatin is characterized by histone modifications (*e.g.* H3–K9 trimethylation and histone tail hypoacetylation), thus linking FRDA to the epigenetic code.[35] This hypothesis is at the basis of one of the most promising therapeutic approaches to FRDA, that based on histone deacetylation inhibitors (see below).

5.4 Frataxin, the Protein Involved in the Disease

Frataxin is a small protein highly conserved in most organisms, from bacteria to primates.[36] It is nearly ubiquitous but with some tissue-specific and developmental expression patterns.[37,38] In adult human tissues, frataxin messenger RNA is most abundant in the spinal cord, heart, skeletal muscle, liver and pancreas, which are the organs most affected. In the central nervous system (CNS), frataxin is highly expressed in the cerebellum and at very low levels in the cerebral cortex.[11] During mouse embryonic development, frataxin starts being detectable in the neuroepithelium around day 10.5. The levels increase substantially at day 14.5 and into the postnatal period.[37,38] In multicellular eukaryotes, frataxin is essential for embryonic development and complete knock-out leads to early embryonic death in plants and mice.[20,39]

The most abundant frataxin transcript (canonical transcript) contains 230 amino acids.[12] Two additional isoforms (isoforms II and III) were recently identified that lack completely (isoform II) or partially (isoform III) the mitochondrial targeting sequence.[12] These isoforms are preferentially localized in different compartments: isoform I is found in mitochondria, whereas isoform II is predominantly located in the cytosol and isoform III in the nucleus. The functional significance of these alternatively spliced and different length isoforms remains, however, unclear.

Isoform I is imported to the mitochondrion thanks to a mitochondrial import signal present in the N-terminus of the transcript Figure 5.4A.[38] The

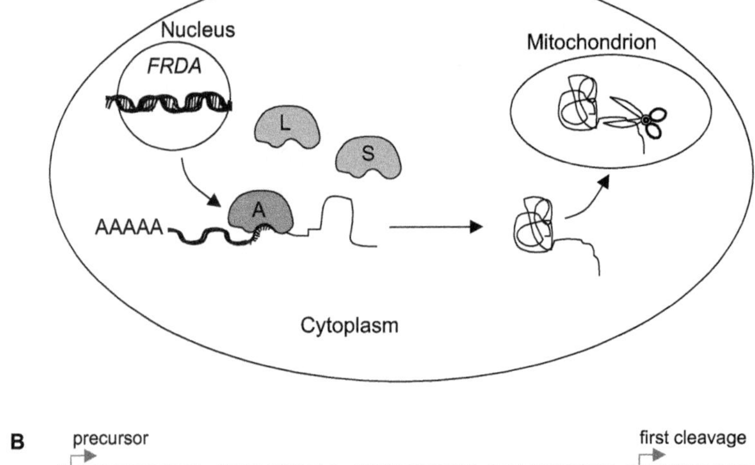

Figure 5.4   Frataxin maturation. (A) The canonical protein is nuclearly encoded, translated in the cytoplasm and then imported to the mitochondria where it is matured. (B) Alternative forms of frataxin. *Arrows* indicate the different isoforms. The mature form spans the sequence 81–210. Alternative forms are 56–210 and 78–210.

precursor is then matured. In yeast and humans, frataxin maturation depends on the mitochondrial processing peptidase (MPP), which cleaves the protein with a two-step cleavage mechanism.[40–43] The mature form of frataxin found in normal individuals and in patients comprises residues 81–210 of the full-length sequence Figure 5.4B.[44] Independent lines of evidence support the view that this form is fully functional and representative of the predominant species in cell. Recombinant frataxin 81–210 co-migrates with endogenous frataxin.[45] It also rescues aconitase activity deficiency[44] and the lethal phenotype in frataxin-deficient murine fibroblasts.[46]

Longer intermediate forms (namely 42–210) can be produced when normal processing is impaired and in some normal cells.[47] *In vitro*, the N-terminus length does not affect the structure[48] or the stability[49] of the protein but has an effect on the biochemical properties.[47] The 81–210 construct is monomeric and binds to iron and to other proteins in this form. The 42–210 construct oligomerizes and forms spheroidal structures that trap iron.

Frataxins are acidic proteins with an isolectric point around pH 4.5. They are highly soluble proteins that, in the absence of heavy cations, exist in solution as monomers.[49,50] The structure of frataxin comprises a C-terminal globular domain conserved in all organisms and an N-terminal region that contains the mitochondrial import signal, which is thus present only in eukaryotes.[36] In human frataxin, the N-terminal signal does not affect the structure or the stability of the globular domain.[48,49] The fold of the globular domain is relatively simple yet almost unique in the protein structure database: it contains two helices sequence-wise flanking an antiparallel β-sheet with 5–7 strands, following an α-(β)$_{5–7}$-α topology (Figure 5.5A).[49,51,52] The two helices pack against the sheet on the same side. This fold is highly conserved, as also expected from sequence conservation.[53] Amongst the factors that stabilize the frataxin fold are divalent and trivalent cations, as well as the length of the region C-terminal to α2.[54] The shorter this region, the more unstable is the protein.

Residues conserved throughout the frataxin family map onto a very specific surface of the protein that comprises helix α1 and the β-sheet (Figure 5.5B).[49,51,52] This region contains, for instance, a semi-conserved acidic region formed of glutamates and aspartates and a completely conserved tryptophan (Trp155 in human frataxin) that is surface exposed on the β-sheet. Clinically important mutations affect two types of residues: some involve mutation of residues buried in the hydrophobic core which will then produce a protein more prone to degradation and/or misfolding.[49,55,56] Others, possibly more interesting, target exposed conserved residues, suggesting that they must be functionally important. Amongst these residues is the conserved Trp155.

5.5 Metals Involved; Pathways Activated by Metals

5.5.1 Frataxin and Metals

Since its first identification, frataxin has been the subject of intense studies aimed at the determination of its cellular role. It is clear that, beyond being

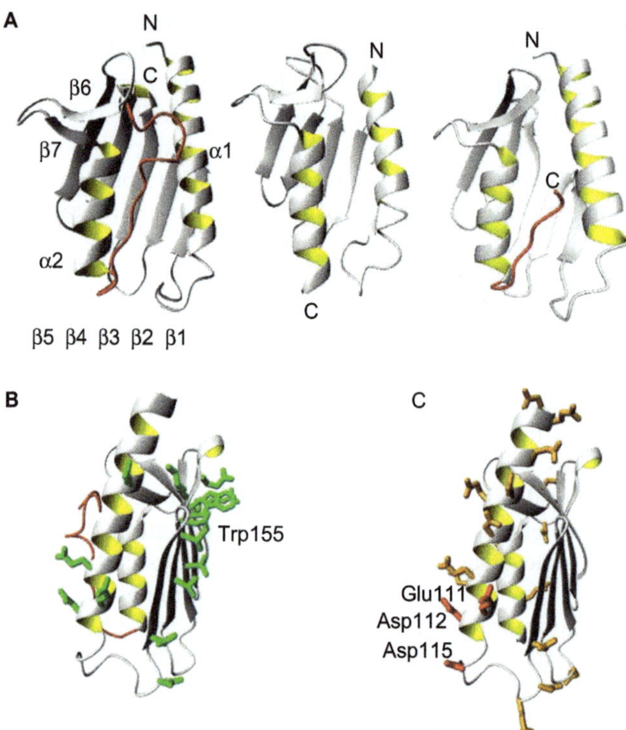

Figure 5.5 Three-dimensional structure of the conserved C-terminal domain. (A) Ribbon representation of the X-ray structures of bacterial (1ew4), yeast (3oeq) and human frataxins (1ekg). Strands are indicated as *arrows*. The secondary structure elements are indicated on the human structure. The N- and C-termini are indicated and the C-termini are coloured in *red*. Their lengths in the three structures are different and correlate with the temperature of thermal unfolding (T_m) of these proteins. (B) The human frataxin structure rotated by 150° along the *y*-axis. Only the backbone atoms are indicated, except for conserved and semi-conserved residues (indicated in *green*). (C) As in B but showing the residues involved in iron binding. All acidic residues are indicated in *orange*. Among them is the triad Glu111, Asp112, Asp115 (in *red*). These residues are conserved and represent the primary iron binding site on monomeric frataxin.

involved in a neurodegenerative process, frataxin is a protein essential for life. Understanding its precise role can thus help us both to design a therapy for FRDA and to clarify an important metabolic pathway. Shortly after the identification of frataxin, it was noticed that iron accumulates in the mito-chondria of *S. cerevisiae* frataxin knock-outs.[57,58] Similar observations were then reported in tissues from FRDA patients.[59] It was at this stage that Isaya and co-workers demonstrated that yeast frataxin, Yfh1, oligomerizes when exposed to excess iron and suggested that frataxins could be ferritin-like proteins.[60] This hypothesis, however, is in conflict with the evidence that oligomerization is favoured only at low ionic strength[50] and is non-essential for the protein function.[61] It is also curious to see that, in higher organisms, mitochondria contain a specialized ferritin that scavenges iron, preventing its accumulation.[62] The presence of this ferritin would make the role of frataxin redundant.

A different hypothesis, which has remained in fashion for quite a long time, is based on the properties of the frataxin monomer and is that frataxin is an iron chaperone. A direct interaction between monomeric frataxin and iron (both Fe^{2+} and Fe^{3+}) was proven experimentally by a combination of calorimetric assays and tryptophan fluorescence.[63] A stoichiometry of *ca.* six cations per frataxin was estimated using the human protein.[64] In support of a direct interaction of iron with the monomer were protein stability studies that showed a strong stabilizing effect of Fe^{2+} and Fe^{3+} cations on the frataxin fold.[54] The affinity is only modest (in the micromolar range).[64] Mapping of the iron binding site was later achieved by NMR studies, which showed that human and bacterial (CyaY) frataxins bind to iron through the semi-conserved acidic surface, supporting a role of this interaction in the frataxin function (Figure 5.5C).[65]

It was also shown, however, that iron is by far not the only cation that frataxin binds.[65,66] Since the interaction solely involves acidic residues, binding is mainly mediated by electrostatic interactions, which are less specific than the coordination by histidine and cysteine residues found in most of the classical iron-binding proteins. As a result, the bacterial frataxin orthologue, CyaY, was shown to bind on the same surface several divalent and trivalent cations other than iron, including copper, magnesium, lutetium, lanthanum and other lanthanides.[65,66]

While certainly suggestive of a link between frataxin and iron metabolism, these findings posed the problem of whether mitochondrial iron accumulation has a causal role in FRDA or is just a consequence of a more direct phenomenon.

A strong piece of evidence in favour of a specific role of metal binding was suggested by Foury and co-workers, who showed that mutation of three acidic residues (Glu111, Asp 112 and Asp115 in human frataxin) involved in the interaction with iron to lysines strongly affects yeast viability.[67]

5.5.2 Frataxin and Iron–Sulphur Cluster Biogenesis

Most of the early evidence pointed towards a link between frataxin and a specific form of iron, *i.e.* iron–sulphur clusters. These are essential prosthetic groups formed by iron and sulphur that provide redox potential to the cell.[68,69] Through binding to specific carrier proteins, iron–sulphur clusters are involved in various processes which include energy metabolism (*e.g.* aconitase and complexes I, II, and III of the respiratory chain), iron metabolism (iron-responsive protein I, ferrochelatase), DNA repair and purine synthesis. Iron–sulphur clusters are formed and assembled into the final acceptor proteins thanks to a tightly regulated complex machinery whose components are highly conserved in most organisms.[69] In prokaryotes these machines are grouped in operons, with the most conserved and widely spread being the ISC operon.

It was noticed that lack of frataxin leads to a deficit of iron–sulphur cluster enzymes in mitochondria and in extra-mitochondrial compartments.[70–73] Approximately at the same time a bioinformatics study demonstrated the co-evolution of frataxin with components of the iron–sulphur cluster biogenesis machine.[74]

Pull-down experiments and biophysical studies independently confirmed this hypothesis and demonstrated a direct interaction between frataxin and the complex of Nfs1 and Isu1 (or IscS and IscU in bacteria, respectively).[75–78] These proteins are the two major and highly conserved components of the machine. Nfs1/IscS is a desulphurase that converts cysteine into alanine and persulphide. Isu/IscU is the scaffold on which the cluster is formed.[79] The two proteins form a complex that is necessary for cluster formation (Figure 5.6). It was then widely assumed that the role of frataxin would be that of an iron transporter which chaperones the metal to be converted into the cluster.[63]

For several years this hypothesis became universally adopted without further investigation. Finally, it was demonstrated that, by forming a stoichiometric ternary complex with the IscS/IscU complex, bacterial frataxin does not just act

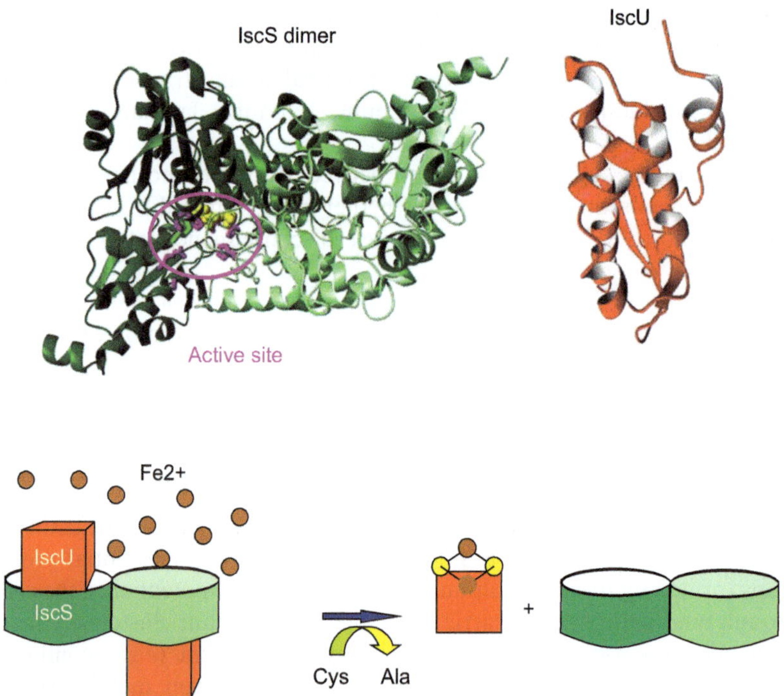

Figure 5.6 The mechanism of iron–sulphur cluster formation. The machine is centred on the desulphurase IscS/Nfs1 and the scaffold protein IscU/Isu. The IscS dimer converts cysteine into alanine and persulphide under anaerobic reducing conditions. The persulphide is transferred on IscU where the cluster is mounted.

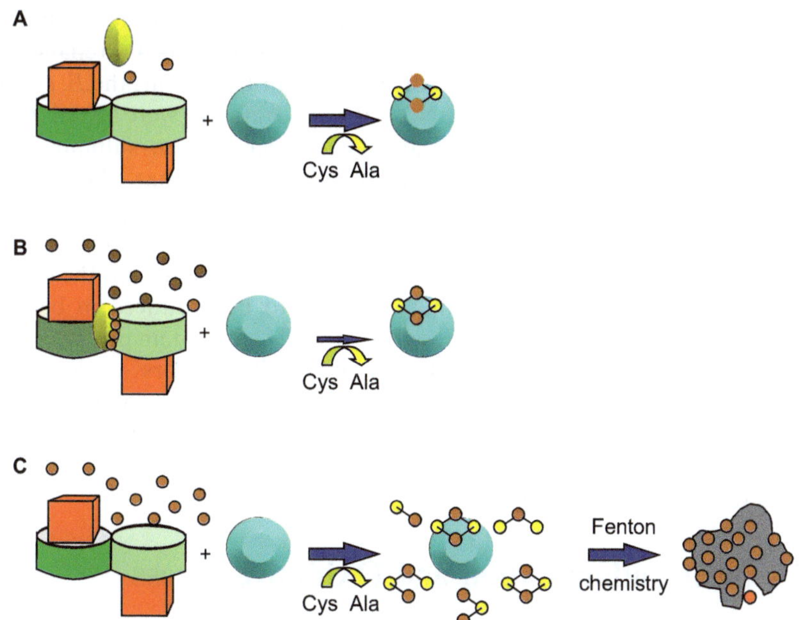

Figure 5.7 Schematic model of the molecular mechanism of frataxin in the cell. (A) At normal iron concentrations, the Fe–S clusters are assembled by the IscS–IscU complex and passed on to their final acceptors (*cyan sphere*). (B) Any excess of iron compared to the number of final acceptors will be balanced by the presence of frataxin (*yellow spheroid*) that slows down the reaction to match the concentration of final acceptors and avoids unnecessary overproduction of Fe–S clusters. (C) When frataxin is absent or produced in insufficient quantities, as in FRDA, there is no regulation. Fe–S clusters will be produced irrespectively of whether they can be transferred to an acceptor. Any iron excess will result in a surplus of Fe–S clusters, which, being highly unstable, will fall apart, generating Fenton reactions. Fe^{3+} will precipitate and form insoluble aggregates (*grey shape*).

as an iron chaperone but is a regulator of the rates of cluster formation (Figure 5.7).[78,80] *In vitro* experiments showed that bacterial frataxin is an inhibitor of the iron–sulphur cluster machinery encoded in the ISC operon, interacting directly with the desulphurase IscS. It was then suggested that frataxin acts as a gatekeeper of iron–sulphur cluster biogenesis by fine tuning the quantity of Fe–S cluster formed to match the concentration of the apo acceptors. In this model, frataxin would have low affinity for the IscS/IscU system at normal iron levels. As soon as there is even a slight iron imbalance, *i.e.* too much iron compared to the amount of final acceptors, the affinity of frataxin for IscS would increase.

These findings were successively confirmed by different techniques[81] and by independent groups who showed that eukaryotic frataxins also play a role in the regulation of iron–sulphur cluster formation.[45,82] It was surprising, however, to see that, while bacterial frataxins slow down cluster formation, eukaryotic frataxins seem to act as activators. To solve this discrepancy and pin down the protein(s) responsible for the different behaviour, the *E. coli* and human proteins (IscS, IscU, CyaY, Nfs1, Isu and frataxin) were mixed *in vitro* using all the possible permutations and their effect on iron–sulphur cluster formation tested.[83] This study conclusively showed that the different behaviour is not due to frataxin but that the responsible protein is the desulphurase Nfs1/IscS; frataxin/CyaY have a reversed behaviour when the desulphurases are swapped. However, a non-trivial point in the correct interpretation of these results is that, in order to be produced in bacteria, recombinant Nfs1 required co-expression with Isd11, a protein present only in eukaryotes.[84,85] More recently, Dancis and collaborators were able to produce Nfs1 in isolation and proved that the enzyme also requires Isd11 to be active.[86] Taken together, these results suggest that regulation of Nfs1 is more complex than previously anticipated and indicate the need for further studies to clarify the role of frataxin.

5.6 Neuroinflammation and Oxidative Stress

An important phenomenon associated with frataxin deficiency is oxidative stress, which is usually related to directly or indirectly perturbed iron metabolism.[87–89] These observations are in good agreement and can be readily explained by the hypothesis that frataxin acts as a gatekeeper of iron–sulphur cluster formation.[78,80] The first report of oxidative stress was observed in frataxin-deficient yeast strains,[57] followed by cultured cells from FRDA patients.[90] Further studies on fibroblasts from FRDA patients confirmed oxidant sensitivity,[91,92] suggesting that even non-affected cell types of FRDA patients may be in a chronic oxidative stress state.

In frataxin-deficient cells there is an increase both in iron and hydrogen peroxide (H_2O_2), which increases as a consequence of the decreased levels of ISC-containing proteins such as the respiratory complexes I, II and III. Peroxide accumulation together with Fe^{2+} generates Fe^{3+} and highly reactive hydroxyl radicals (the Fenton reaction).[93–97] The presence of oxidative stress in the urine and blood of FRDA patients[98–100] was directly supported by different biomarkers, such as 8-hydroxy-2′-deoxyguanosine (a marker of DNA oxidative damage), plasma malondialdehyde (a lipid peroxidation product) and plasma glutathione-*S*-transferase activity.[101] Additionally, different signs of altered glutathione homeostasis were reported.[95,102,103] It was recently suggested that an impaired response of antioxidant enzymes could also be the consequence of misfunction of the Nrf2-dependent phase II antioxidant-signalling pathway.[104]

It is, however, worth mentioning that, while most of the evidence supports oxidative stress, these reports have not remained unchallenged; other findings could argue against the presence of oxidative stress. No sign of oxidative damage was detected even in cases of advanced disease in the heart muscle of conditional knock-out mice.[105] Analysis of oxidative stress markers in urine from FRDA patients also showed some inconsistencies.[106,107] A possible explanation for the latter findings could be that biomarkers are only statistical indicators of a cellular state and that the levels of biomarkers can fluctuate with time. It is also possible that the complete shutdown of the respiratory chain that is expected in conditional knock-out mice when frataxin is completely absent could totally prevent oxidative stress.

5.7 Therapeutic Strategies

As with most of the neurodegenerative diseases, FRDA is currently incurable. Identification of the responsible gene and determination of the function of frataxin could, however, hold a promise that treatment can be developed in a non-too-distant future. Different directions are being explored, some of which have been shown to be inconclusive, others being more promising. Two main problems are encountered in the rigorous evaluation of the proposed treatments. The patient cohorts are often too small to be statistically meaningful. Ethical reasons have also prevented the use of double-blind randomized placebo-controlled studies in some countries. While certainly understandable, it is difficult to have a clear picture of the effects of the treatment without a control. In the next sections, some of the approaches undertaken will be briefly described.

5.7.1 Antioxidants

The link between frataxin, iron and oxidative damage readily suggested the use of antioxidants as an obvious therapeutic approach to FRDA. Amongst these compounds, the use of quinone antioxidants, such as CoQ and vitamin E derivatives, was explored first. A pilot study using a combination of CoQ_{10} and vitamin E suggested some improvement of mitochondrial function and reduction of oxidative stress, with some beneficial effects on bioenergetic aspects of cardiac and skeletal muscle.[108]

Idebenone, a short-chain analogue of CoQ, was then proposed as an alternative because it is supposed to enter mitochondria more efficiently than CoQ_{10} (although it is difficult to track down solid evidence in support of this statement). It was argued that idebenone reduces membrane lipid peroxidation and rescues CoQ-deficient cells.[109] The first open non-placebo idebenone trial reported a decrease in heart size in three out of ten FRDA patients with hypertrophic cardiomyopathy.[110] However, the effect was not confirmed in other controlled clinical trials.[109,111,112] All randomized controlled trials using idebenone have shown no significant benefit on neurological symptoms.[109,112–115] It has been argued that while idebenone does not affect disease progression and/or the neurologic status in patients compared to placebo, it may still have some beneficial effects for cardiomyopathy and reduce fatigue status, at least in some patients. Given the high cost, however, the treatment does not seem to have a promising cost-to-benefit ratio.

5.7.2 Iron Chelators

Another obvious approach to FRDA treatment is the use of iron chelators: small, high-affinity compounds that can scavenge Fe^{3+} and reduce the toxic effects of free radicals.[116,117] An intensively studied iron chelator is deferiprone, a lipid-soluble 1,2-diketo heterocycle. This compound used in FRDA mRNA-silenced HEK-293 cells was shown to chelate mitochondrial iron, leading to restoration of mitochondrial membrane redox potential, increase of ATP production and resistance to apoptosis.[118] In the same study the effect of other siderophores such as deferasirox, salicylaldehyde isonicotinoyl hydrazone and deferoxamine was also compared. It was suggested that the efficacies of these compounds in HEK-293 cells could inversely correlate with iron affinity[118] and that a lower affinity for iron might result in a more efficacious treatment. However, since the affinities of the compounds are practically the same within experimental error, the observed effect must most likely reflect other factors. Despite the initial promise, deferiprone treatment has a number of negative aspects. In some patients it was observed to induce neutropenia, which is characterized by an abnormally low number of neutrophils. Excessive iron chelation might also be beneficial only at some stage of the disease, but the danger is to reduce the mitochondrial iron level too much.[119] The use of this drug needs thus to be further established.

5.7.3 Epigenetic Approaches

As previously mentioned, decreased expression of frataxin correlates with hypoacetylation of histones H3 and H4 and trimethylation of histone H3, suggesting a heterochromatin-mediated repression mechanism.[120] These observations suggested the possibility that histone deacetylase (HDAC) inhibitors could promote a transition from silent heterochromatin to an active chromatin conformation and thus be effective in increasing frataxin expression. A similar approach has been adopted for other neurodegenerative diseases caused by misfolding and in cancer treatments.[121] The main problem with this approach could be, however, that most of the HDAC inhibitors are toxic. A way around this limitation was found by Gottesfeld at Scripps, who managed to develop a new family of HDAC inhibitors that increase frataxin expression without acute toxicity.[120] These are all pimelic diphenylamide derivatives that directly interact with class I HDACs (particularly HDAC3) by reverting histones H3 and H4 hypoacetylation in the chromatin regions upstream and downstream from the GAA repeat, without affecting H3K9 trimethylation that is the hallmark of heterochromatin.[120,122] More recently, a new compound was designed and tested in FRDA patients' lymphocytes and in mouse models. It seems to have the most effective properties for up-regulating frataxin expression with pharmacological effects on histone acetylation and frataxin levels that largely outlast the persistence of the molecule in the tissue.[123] A phase 1 clinical trial of the compound was started in Italy last autumn under the control of the RepliGen Corporation (Waltham, Massachusetts).

This approach is clearly the most promising amongst the present possibilities.

5.7.4 TAT-frataxin

A different way to increase the levels of frataxin could in principle be to directly supplement frataxin in patients, provided that it could be efficiently imported in the cellular environment. Vyas and colleagues found that a fusion of frataxin with the protein transduction domain of the TAT protein (TAT-frataxin) successfully delivers frataxin to mitochondria in cultured cells from FRDA patients and from a mouse model.[124] TAT (transactivator of transcription) from the human immunodeficiency virus is a short peptide able to deliver fused proteins across cellular and intracellular membranes. TAT fusion proteins containing a mitochondrial targeting sequence can translocate through the mitochondrial membranes, with appropriate processing and persistence of the fusion protein within mitochondria.[125,126] Injection of TAT-frataxin into a conditional-knockout mouse model improved growth, increased lifespan and cardiac function.[124] This approach is interesting even though many questions remain about the real application in patients.

5.7.5 Other Approaches

In addition to the main strategies, several other approaches have been explored. Recombinant human erythropoietin, a glyco-protein hormone that controls red blood cell production and acts as a cytokine (protein signalling molecule) precursor in bone marrow, was shown to increase frataxin expression.[127] Since the hormone does not lead to frataxin mRNA increase, the effect is thought to be the consequence of posttranscriptional events.[128,129] Erythropoietin mimetics, which might increase frataxin expression without a concomitant increase in haematocrit, are also under investigation.

A different approach attempted by Testi and co-workers is inhibition of the mechanisms that regulate frataxin stability and degradation in cell. They showed that human frataxin Lys147 is the critical residue responsible for frataxin ubiquitination and degradation by the ubiquitin-proteasome system.[130] Effective identification of compounds that could prevent frataxin ubiquitination and degradation could thus constitute an interesting possibility.

Other strategies to increase frataxin include high-throughput drug screenings using a frataxin reporter system to identify compounds that increase frataxin expression, and resveratrol, which increases frataxin expression in multiple *in vitro* FRDA models.[131]

5.8 Conclusions

Discovery of the frataxin gene has been a real leap forward that was essential before even considering the possibility of FRDA treatment. Yet, almost two decades later, the mechanisms that determine production, silencing and function(s) of frataxin are still unclear. What is now established is that frataxin is not only a protein associated with a tragic rare disease but also an essential component of the cell whose absence is incompatible with life. Accordingly, solid evidence links the frataxin function to iron–sulphur cluster biogenesis, a metabolic pathway essential for survival and about which our understanding is rapidly growing. We can thus hope that an increasing number of scientists will want to take on the challenge of clarifying how this complex and yet fascinating machinery works. Meanwhile, there are an increasing number of therapeutic approaches that seem promising and well focused. Among them are, of particular interest, the HDAC inhibitors and the TAT fusion protein. Together or as alternative possibilities they could really contribute to the treatment of FRDA.

References

1. G. E. Berrios, *Int. J. Geriatr. Psych.*, 1991, **5**, 355.
2. N. Friedreich, *Virchows Arch. Pathol. Anat.*, 1863, **27**, 1.
3. G. Geoffroy, A. Barbeau, G. Breton, *et al.*, *Can. J. Neurol. Sci.*, 1976, **3**, 279.
4. A. E. Harding, *Brain*, 1981, **104**, 589.
5. A. E. Harding and K. J. Zilkha, *J. Med. Genet.*, 1981, **18**, 285.
6. M. Pandolfo, *J. Neurol.*, 2009, **256**(suppl. 1), 3.
7. M. Cossée, M. Schmitt, V. Campuzano, L. Reutenauer, C. Moutou, J. L. Mandel and M. Koenig., *Proc. Natl. Acad. Sci. U. S. A.*, 1997, **94**, 7452.
8. M. L. Moseley, K. A. Benzow, L. J. Schut, *et al.*, *Neurology*, 1998, **51**, 1666.
9. S. Chamberlain, J. Shaw, A. Rowland, *et al.*, *Nature*, 1988, **334**, 248.
10. S. Chamberlain, M. Farrall, J. Shaw, *et al.*, *Am. J. Hum. Genet.*, 1993, **52**, 99.
11. V. Campuzano, L. Montermini, M. D. Molto, L. Pianese, M. Cossée, F. Cavalcanti, E. Monros, F. Rodius, F. Duclos, A. Monticelli, *et al.*, *Science*, 1996, **271**, 1423.
12. H. Xia, Y. Cao, X. Dai, Z. Marelja, D. Zhou, R. Mo, S. Al-Mahdawi, M. A. Pook, S. Leimkuhler, T. A. Rouault and K. Li, *PloS One*, 2012, **7**, e47847.
13. A. Filla, G. De Michele, F. Cavalcanti, *et al.*, *Am. J. Hum. Genet.*, 1996, **59**, 554.
14. A. Dürr, M. Cossée, Y. Agid, *et al.*, *New Engl. J. Med.*, 1996, **335**, 1169.
15. R. Sharma, S. Bhatti, M. Gomez, *et al.*, *Hum. Mol. Genet.*, 2002, **11**, 2175.
16. L. Pollard, R. Sharma, M. Gomez, *et al.*, *Nucleic Acids Res.*, 2004, **32**, 5962.
17. R. Sharma, I. De Biase, M. Gómez, *et al.*, *Ann. Neurol.*, 2004, **56**, 898.

18. J. R. Gatchel and H. Y. Zoghbi, *Nat. Rev. Genet.*, 2005, **6**, 743 .
19. A. Sailer and H. Houlden, *Curr. Neurol. Neurosci. Rep.*, 2012, **12**, 227.
20. M. Cossée, H. Puccio and A. Gansmuller, *et al.*, *Hum. Mol. Genet.*, 2000, **9**, 1219.
21. M. Cossée, A. Durr, M. Schmitt, N. Dahl, P. Trouillas, P. Allinson, M. Kostrzewa, A. Nivelon-Chevallier, K. H. Gustavson, A. Kohlschutter, *et al.*, *Ann. Neurol.*, 1999, **45**, 200.
22. C. Gellera, B. Castellotti, C. Mariotti, R. Mineri, V. Seveso, S. Didonato and F. Taroni, *Neurogenetics*, 2007, **8**, 289.
23. L. M. Luheshi and C. M. Dobson, *FEBS Lett.*, 2009, **583**, 2581.
24. M. Vendruscolo, T. P. Knowles and C. M. Dobson, *Cold Spring Harb. Perspect. Biol.*, 2011, **3**, a010454.
25. C. M. Gomes, *Curr. Top. Med. Chem.*, 2013, in press.
26. K. Ohshima, L. Montermini, R. D. Wells and M. Pandolfo, *J. Biol. Chem.*, 1998, **273**, 14588.
27. R. D. Wells, *FASEB J.*, 2008, **22**, 1625.
28. S. V. Mariappan, P. Catasti, L. A. Silks 3rd, E. M. Bradbury and G. Gupta, *J. Mol. Biol.*, 1999, **285**, 2035.
29. N. Sakamoto, P. D. Chastain, P. Parniewski, K. Ohshima, M. Pandolfo, J. D. Griffith and R. D. Wells, *Mol. Cell*, 1999, **3**, 465.
30. N. Sakamoto, K. Ohshima, L. Montermini, M. Pandolfo and R. D. Wells, *J. Biol. Chem.*, 2001, **276**, 27171.
31. R. D. Wells, R. Dere, M. L. Hebert, M. Napierala and L. S. Son, *Nucleic Acids Res.*, 2005, **33**, 3785.
32. A. A. Vetcher, M. Napierala, R. R. Iyer, P. D. Chastain, J. D. Griffith and R. D. Wells, *J. Biol. Chem.*, 2002, **277**, 39217.
33. L. S. Son and R. D. Wells, in *Genetic Instabilities and Hereditary Neurological Diseases*, ed. R. D. Wells and T. Ashizawa, Academic/ Elsevier, San Diego, pp. 327–335.
34. A. Saveliev, C. Everett and T. Sharpe, *et al.*, *Nature*, 2003, **422**, 909.
35. S. C. Elgin and S. I. Grewal, *Curr. Biol.*, 2003, **13**, R895.
36. T. J. Gibson, E. V. Koonin, G. Musco, A. Pastore and P. Bork, *Trends Neurosci.*, 1996, **19**, 465.
37. S. Jiralerspong, Y. Liu, L. Montermini, S. Stifani and M. Pandolfo, *Neurobiol. Dis.*, 1997, **4**, 103.
38. H. Koutnikova, V. Campuzano, F. Foury, P. Dollé, O. Cazzalini and M. Koenig, *Nat. Genet.*, 1997, **16**, 345.
39. H. Puccio, *J. Neurol.*, 2009, **256**(suppl. 1), 18.
40. H. Koutnikova, V. Campuzano and M. Koenig, *Hum. Mol. Genet.*, 1998, **7**, 1485.
41. S. S. Branda, P. Cavadini, J. Adamec, F. Kalousek, F. Taroni and G. Isaya, *J. Biol. Chem.*, 1999, **274**, 22763.
42. P. Cavadini, J. Adamec, F. Taroni, O. Gakh and G. Isaya, *J. Biol. Chem.*, 2000, **275**, 41469.
43. D. Gordon, M. Kogan, S. Knight, *et al.*, *Hum. Mol. Genet.*, 2001, **10**, 259.

44. I. Condò, N. Ventura, F. Malisan, A. Ruffini, B. Tomassini and R. Testi, *Hum. Mol. Genet.*, 2007, **16**, 1534.

45. S. Schmucker, A. Martelli, F. Colin, A. Page, M. Wattenhofer-Donze, L. Reutenauer and H. Puccio, *PLoS One*, 2011, **6**, e16199.

46. S. Schmucker, M. Argentini, N. Carelle-Calmels, A. Martelli and H. Puccio, *Hum. Mol. Genet.*, 2008, **17**, 3521.

47. O. Gakh, T. Bedekovics, S. F. Duncan, D. Y. Smith IV, D. S. Berkholz and G. Isaya, *J. Biol. Chem.*, 2010, **285**, 38486.

48. F. Prischi, C. Giannini, S. Adinolfi and A. Pastore, *FEBS J.*, 2009, **276**, 6669.

49. G. Musco, G. Stier, B. Kolmerer, S. Adinolfi, S. Martin, T. Frenkiel, T. Gibson and A. Pastore, *Struct. Fold. Des.*, 2000, **8**, 695.

50. S. Adinolfi, M. Trifuoggi, A. Politou, S. Martin and A. Pastore, *Hum. Mol. Genet.*, 2002, **11**, 1865.

51. S. Dhe-Paganon, R. Shigeta, Y. I. Chi, M. Ristow and S. E. Shoelson, *J. Biol. Chem.*, 2000, **275**, 30753.

52. S. J. Cho, M. G. Lee, J. K. Yang, J. Y. Lee, H. K. Song and S. W. Suh, *Proc. Natl. Acad. Sci. U. S. A.*, 2000, **97**, 8932.

53. Y. He, S. L. Alam, S. V. Proteasa, Y. Zhang, E. Lesuisse, A. Dancis and T. L. Stemmler, *Biochemistry*, 2004, **43**, 16254.

54. S. Adinolfi, M. Nair, A. Politou, E. Bayer, S. Martin, P. A. Temussi and A. Pastore, *Biochemistry*, 2004, **43**, 6511.

55. A. R. Correia, S. Adinolfi, A. Pastore and C. Gomes, *Biochem. J.*, 2006, **398**, 605.

56. A. R. Correia, C. Pastore, S. Adinolfi, A. Pastore and C. M. Gomes, *FEBS J.*, 2008, **275**, 3680.

57. M. Babcock, D. de Silva, R. Oaks, S. Davis-Kaplan, S. Jiralerspong, L. Montermini, M. Pandolfo and J. Kaplan, *Science*, 1997, **276**, 1709.

58. F. Foury and O. Cazzalini, *FEBS Lett.*, 1997, **411**, 373.

59. J. L. Bradley, J. C. Blake, S. Chamberlain, P. K. Thomas, J. M. Cooper and A. H. Schapira, *Hum. Mol. Genet.*, 2000, **9**, 275.

60. J. Adamec, F. Rusnak, W. G. Owen, S. Naylor, L. M. Benson, A. M. Gacy and G. Isaya, *Am. J. Hum. Genet.*, 2000, **67**, 549.

61. K. Aloria, B. Schilke, A. Andrew and E. A. Craig, *EMBO Rep.*, 2004, **5**, 1096.

62. S. Levi, B. Corsi, M. Bosisio, R. Invernizzi, A. Volz, D. Sanford, P. Arosio and J. Drysdale, *J. Biol. Chem.*, 2001, **276**, 24437.

63. T. Yoon and J. A. Cowan, *J. Am. Chem. Soc.*, 2003, **125**, 6078.

64. F. Bou-Abdallah, S. Adinolfi, A. Pastore, T. M. Laue and N. D. Chasteen, *J. Mol. Biol.*, 2004, **341**, 605.

65. M. Nair, S. Adinolfi, C. Pastore, G. Kelly, P. Temussi and A. Pastore, *Structure*, 2004, **12**, 2037.

66. C. Pastore, M. Franzese, F. Sica, P. A. Temussi and A. Pastore, *FEBS J.*, 2007, **274**, 4199.

67. F. Foury, A. Pastore and M. Trincal, *EMBO Rep.*, 2007, **8**, 194.

68. D. J. Johnson, D. R. Dean, A. D. Smith and M. K. Johnson, *Annu. Rev. Biochem.*, 2005, **74**, 247.
69. B. Py and F. Barras, *Nat. Rev.*, 2010, **8**, 437.
70. J. B. Lamarche, D. Shapcott, M. Côté, *et al.*., in *Handbook of Cerebellar Diseases*, ed. R. Lechtenberg, Dekker, New York, 1993, pp. 453–458.
71. A. Rotig, P. de Lonlay, D. Chretien, F. Foury, M. Koenig, D. Sidi, A. Munnich and P. Rustin, *Nat. Genet.*, 1997, **17**, 215.
72. G. Duby, F. Foury, A. Ramazzotti, J. Herrmann and T. Lutz, *Hum. Mol. Genet.*, 2002, **11**, 2635.
73. H. Puccio, D. Simon, M. Cossée, P. Criqui-Filipe, F. Tiziano, J. Melki, C. Hindelang, R. Matyas, P. Rustin and M. Koenig, *Nat Genet.*, 2001, **27**, 181.
74. M. A. Huynen, B. Snel, P. Bork and T. J. Gibson, *Hum. Mol. Genet.*, 2001, **10**, 2463.
75. A. Ramazzotti, V. Vanmansart and F. Foury, *FEBS Lett.*, 2004, **557**, 215.
76. J. Gerber, U. Muhlenhoff and R. Lill, *EMBO Rep.*, 2003, **4**, 906.
77. G. Layer, S. Ollagnier de Choudens, Y. Sanakis and M. Fontecave, *J. Biol. Chem.*, 2006, **281**, 16256.
78. S. Adinolfi, C. Iannuzzi, F. Prischi, C. Pastore, S. Iametti, S. R. Martin, F. Bonomi and A. Pastore, *Nat. Struct. Mol. Biol.*, 2009, **16**, 390.
79. R. Shi, A. Proteau, M. Villarroya, I. Moukadiri, L. Zhang, J. F. Trempe, A. Matte, M. E. Armengod and M. Cygler, *PLoS Biol.*, 2010, **8**, e1000354.
80. F. Prischi, P. V. Konarev, C. Iannuzzi, C. Pastore, S. Adinolfi, S. R. Martin, D. I. Svergun and A. Pastore, *Nat. Commun*, 2010, **1**, 5.
81. C. Iannuzzi, S. Adinolfi, B. D. Howes, R. Garcia-Serres, M. Clémancey, J. M. Latour, G. Smulevich and A. Pastore, *PloS One*, 2011, **6**, e21992.
82. C. I. Tsai and D. P. Barondeau, *Biochemistry*, 2010, **49**, 9132.
83. J. Bridwell-Rabb, C. Iannuzzi, A. Pastore and D. P. Barondeau, *Biochemistry*, 2012, **51**, 2506.
84. N. Wiedemann, E. Urzica, B. Guiard, H. Müller, C. Lohaus, H. E. Meyer, M. T. Ryan, C. Meisinger, U. Mühlenhoff, R. Lill and N. Pfanner, *EMBO J.*, 2006, **25**, 184.
85. A. C. Adam, C. Bornhövd, H. Prokisch, W. Neupert and K. Hell, *EMBO J.*, 2006, **25**, 174.
86. A. Pandey, R. Golla, H. Yoon, A. Dencis and D. Pain, *Biochem. J.*, 2012, **448**, 171.
87. S. Al-Mahdawi, R. M. Pinto, D. Varshney, *et al.*, *Genomics*, 2006, **88**, 580.
88. M. V. Busi, M. V. Maliandi, H. Valdez, *et al.*, *Plant J.*, 2006, **48**, 873.
89. R. Vázquez-Manrique, P. González-Cabo, S. Ros, *et al.*, *FASEB J.*, 2006, **20**, 172.
90. A. Wong, J. Yang, P. Cavadini, C. Gellera, B. Lonnerdal, F. Taroni and G. Cortopassi, *Hum. Mol. Genet.*, 1999, **8**, 425.
91. K. Chantrel-Groussard, V. Geromel, H. Puccio, *et al.*, *Hum. Mol. Genet*, 2001, **10**, 2061.

92. S. Jiralerspong, B. Ge, T. J. Hudson, *et al.*, *FEBS Lett.*, 2001, **509**, 101.

93. M. Santos, K. Ohshima and M. Pandolfo, *Hum. Mol. Genet.*, 2001, **10**, 1935.

94. R. Santos, N. Buisson, S. A. Knight, *et al.*, *Mol. Microbiol.*, 2004, **54**, 507.

95. E. Napoli, F. Taroni and G. Cortopassi, *Antioxid. Redox Signal.*, 2006, **8**, 506.

96. V. Irazusta, E. Cabiscol, G. Reverter-Branchat, *et al.*, *J. Biol. Chem.*, 2006, **281**, 12227.

97. A. Seguin, A. Bayot, A. Dancis, A. Rogowska-Wrzesinska, F. Auchère, J. M. Camadro, A. L. Bulteau and E. Lesuisse, *Mitochondrion*, 2009, **9**, 130.

98. M. Emond, G. Lepage, M. Vanasse, *et al.*, *Neurology*, 2000, **55**, 1752.

99. J. L. Bradley, S. Homayoun, P. E. Hart, *et al.*, *Neurochem. Res.*, 2004, **29**, 561.

100. J. Schulz, T. Dehmer, L. Schöls, *et al.*, *Neurology*, 2000, **55**, 1719.

101. G. Tozzi, M. Nuccetelli, M. Lo Bello, S. Bernardini, L. Bellincampi, S. Ballerini, L. M. Gaeta, C. Casali, A. Pastore, G. Federici, E. Bertini and F. Piemonte, *Arch. Dis. Child.*, 2002, **86**, 376.

102. F. Auchère, R. Santos, S. Planamente, *et al.*, *Hum. Mol. Genet.*, 2008, **17**, 2790.

103. F. Piemonte, A. Pastore, G. Tozzi, *et al.*, *Eur. J. Clin. Invest.*, 2001, **31**, 1007.

104. V. Paupe, E. Dassa, S. Goncalves, *et al.*, *PLoS One*, 2009, **4**, e4253.

105. H. Seznec, D. Simon, C. Bouton, *et al.*, *Hum. Mol. Genet.*, 2004, **14**, 463.

106. N. Di Prospero, A. Baker, N. Jeffries, *et al.*, *Lancet Neurol.*, 2007, **6**, 878.

107. L. Myers, D. Lynch, J. Farmer, *et al.*, *Mov. Disord.*, 2008, **23**, 1920.

108. P. Hart, R. Lodi, B. Rajagopalan, *et al.*, *Arch. Neurol.*, 2005, **62**, 621.

109. T. Meier and G. Buyse, *J. Neurol.*, 2009, **256**, 25.

110. P. Rustin, J. C. von Kleist-Retzow, K. Chantrel-Groussard, *et al.*, *Lancet*, 1999, **354**, 477.

111. R. Artuch, A. Aracil, A. Mas, *et al.*, *Neuropediatrics*, 2002, **33**, 190.

112. D. Lynch, S. Perlman and T. Meier, *Arch. Neurol.*, 2010, **67**, 941.

113. S. Lagedrost, M. Sutton, M. Cohen, *et al.*, *Am. Heart J.*, 2011, **161**, 639.

114. C. Mariotti, A. Solari, D. Torta, *et al.*, *Neurology*, 2003, **60**, 1676.

115. M. Kearney, R. Orrell, M. Fahey, *et al.*, *Cochrane Database Syst. Rev.*, 2012 (4), art. no. CD007791; doi: 10.1002/14651858.

116. G. J. Kontoghiorghes, A. Efstathiou, M. Kleanthous, Y. Michaelides and A. Kolnagou, *Hemoglobin*, 2009, **33**, 386.

117. D. R. Richardson, *Expert Opin. Invest. Drugs*, 2003, **12**, 235.

118. O. Kakhlon, H. Manning, W. Breuer, N. Melamed-Book, C. Lu, G. Cortopassi, A. Munnich and Z. I. Cabantchik, *Blood*, 2008, **112**, 5219.

119. S. Goncalves, V. Paupe, E. P. Dassa, *et al.*, *BMC Neurol.*, 2008, **8**, 20.

120. D. Herman, K. Jenssen, R. Burnett, *et al.*, *Nat. Chem. Biol.*, 2006, **2**, 551.

121. M. Rai, E. Soragni, C. J. Chou, *et al.*, *PLoS One*, 2010, **5**, e8825.

122. M. Rai, E. Soragni, K. Jenssen, *et al.*, *PLoS One*, 2008, **3**, e1958.
123. M. D. Cantley and D. R. Haynes, *Inflammopharmacology*, 2013, in press.
124. P. M. Vyas, W. J. Tomamichel, P. M. Pride, *et al.*, *Hum. Mol. Genet.*, 2012, **21**, 1230.
125. V. Del Gaizo and R. Payne, *Mol. Ther.*, 2003, **7**, 720.
126. V. Del Gaizo, J. MacKenzie and R. Payne, *Mol. Genet. Metab.*, 2003, **80**, 170.
127. B. Sturm, D. Stupphann, C. Kaun, *et al.*, *Eur. J. Clin. Invest.*, 2005, **35**, 711.
128. F. Acquaviva, I. Castaldo, A. Filla, *et al.*, *Cerebellum*, 2008, **7**, 360.
129. F. Sacca, G. Puorro, A. Antenora, *et al.*, *PLoS One*, 2011, **6**, e17627.
130. A. Rufini, S. Fortuni, G. Arcuri, I. Condò, D. Serio, O. Incani, F. Malisan, N. Ventura and R. Testi, *Hum. Mol. Genet.*, 2011, **20**, 1253.
131. J. P. Sarsero, L. Li, H. Wardan, K. Sitte, R. Williamson and P. A. Ioannou, *J. Gene Med.*, 2003, **5**, 72.

Prion Disease

MAGDALENA ROWINSKA-ZYREK,[a]
DANIELA VALENSIN,[b] MAREK LUCZKOWSKI[a] AND
HENRYK KOZLOWSKI*[a]

[a] Department of Chemistry, University of Wroclaw H, F. Joliot-Curie 14,
50-383 Wroclaw, Poland; [b] Department of Biotechnology, Chemistry and
Pharmacy, University of Siena, via Alcide de Gasperi 2, 53100 Siena, Italy
*Email: henrykoz@wchuwr.chem.uni.wroc.pl

6.1 Definition of the Disease and Clinical Presentation

Prion diseases (derived either from infection, germline mutations or most often occurring sporadically), both in humans and animals, are fatal neuro-degenerative disorders characterized by progressive brain degeneration. It is widely accepted that they are caused by protein-only infectious agents propagating disease by inducing protein conformational changes.[1] The molecular mechanism of prion pathologies is not yet entirely understood but some aspects seem to be generally accepted, such as spongiform degeneration, non-classical inflammation of the brain, progressive neuron loss, accumulation of protein aggregates and synaptic alterations. In humans, the group of prion diseases includes Creutzfeldt–Jakob (CJD), Gerstmann–Sträussler–Scheinker (GSS), fatal familial insomnia (FFI) and kuru diseases.[1-4] In animals the transmissible spongiform encephalopathies (TSEs) include scrapie in sheep, bovine spongiform encephalopathy and chronic wasting disease in deer and elk. The most common prion diseases originate sporadically, but the unusual transmissible (infectious) form has been the most attractive for research. In the case of infection, the disease can be transmitted by several mechanisms, such as blood

RSC Metallobiology Series No. 1
Mechanisms and Metal Involvement in Neurodegenerative Diseases
Edited by Roberta Ward, David Dexter and Robert Crichton
© The Royal Society of Chemistry 2013
Published by the Royal Society of Chemistry, www.rsc.org

transfusion, ingestion and iatrogenic transmission, and it can occur either within or between species. Prion, the infectious agent (PrP^{Sc}, Sc for scrapie), discovered by Prusiner,[5] was eventually found to be a misfolded protein existing mostly in brain, which was the isoform of a natively folded cellular prion protein (PrP^C). PrP^C is a monomeric water-soluble protein consisting of three helical domains and two short β-sheets, while its scrapie form possesses large β-domains and it easily forms protein aggregates.[6,7]

PrP^C is synthesized in the endoplasmic reticulum, processed in the Golgi apparatus and finally transported to the cell membranes, where it is usually located in lipid rafts, to which it is attached *via* a glycosylphosphatidylinositol anchor.[8] Prp^C is a highly conserved protein in mammals, although its biological functions are not yet well established. Several possible physiological activities of PrP^C have been suggested, such as involvement in signalling resulting in neuroprotection,[9,10] antioxidative stress,[11] copper binding[12,13] or protection against cell death.[14]

There is general agreement that formation and accumulation of PrP^{Sc} is a major factor in the development of neurodegeneration and disease, although the relation between the scrapie prions and pathology development is still unclear. One of the possible reasons for the pathology could be the loss of PrP^C biological activity when transformed into the PrP^{Sc} isoform. This hypothesis is still controversial[15] and a more attractive one is the mechanism proposed for other neuropathologies such as Alzheimer's or Parkinson's disease, namely that the major factor is the toxicity of the aggregated PrP^{Sc} itself.[16] Although in most cases this hypothesis seems reasonable, there are cases in which it does not work.[17,18] Thus, the transformation of PrP^C into the PrP^{Sc} isoform is a basic requirement for prion-related diseases, although to date the relation between prions and neurodegeneration is not understood.[19]

It was generally believed that PrP^{Sc} is insoluble and protease resistant. However, small PrP^{Sc} polymers (below 600 kD) can be sensitive to proteolysis.[20] The most infectious prion particles were found to be those of MW 300–600 kD.[21] Thus, the infectious PrP^{Sc} might not be the protease-resistant form, but also the protease-sensitive form as described for a human case.[22]

Experiments carried out with synthetic prions that had been misfolded and aggregated *in vitro*, which induced neurodegenerative disease when injected into transgenic mice, demonstrated the "protein only" hypothesis quite convincingly.[23–25] However, there are still several very critical questions to be answered before prion diseases will be really understood, including: is PrP^{Sc} the only component of the infectious prion agent? How does it induce neurodegenerative pathology? What is the real structure of aggregated PrP^{Sc}?[1] Do metal ions play any critical role in prion-induced neurodegeneration, *e.g.* by having an impact on aggregation or oxidative stress?[26,27]

In the case of human diseases, CJD is usually defined as spongiform encephalopathy, GSS as a encephalo(myelo)pathy with multicentric amyloid plaques and FFI is dominantly a thalamic degeneration.[28] Clinical presentations of different diseases usually differ from each other, although some of the symptoms could be very similar. Progressive dementia, extra-pyramidal and pyramidal dysfunction, visual and cerebellar disturbances are characteristic for sporadic cases of CJD, while in the case of GSS, slowly progressing ataxia and cognitive decline are frequently associated with pseudobulbar palsy.[29,30] Also the duration of the illness could differ for particular prion diseases between several months and several years.

6.2 Genetic Contribution

Human prion diseases comprise three major entities with variable phenotypic features: CJD, FFI and GSS. From an etiological perspective they can be divided into inherited, sporadic and acquired forms. The sporadic form constitutes approximately 85% of all CJD cases and occurs annually at the rate of 1–2 patients per million of the population worldwide. Around 15% of cases are associated with autosomal dominant mutations in the *PRNP* gene, while the remaining rare cases fall into the category of acquired human prion diseases. So far, over 30 mutations linked to chromosome 20, mostly targeting the *PRNP* gene, are known to determine familiar cases of the disease.[31–36] They include premature stop codons, amino acid substitutions and insertions of octapeptide repeats (Figure 6.1).[37] In fact, two major entities of prion diseases, genetic CJD and FFI, are linked to two types of mutations of the *PRNP* gene, point mutations leading to amino acid substitutions and deletion/insertion mutations yielding PrP with an altered number of octapeptide repeats.[38]

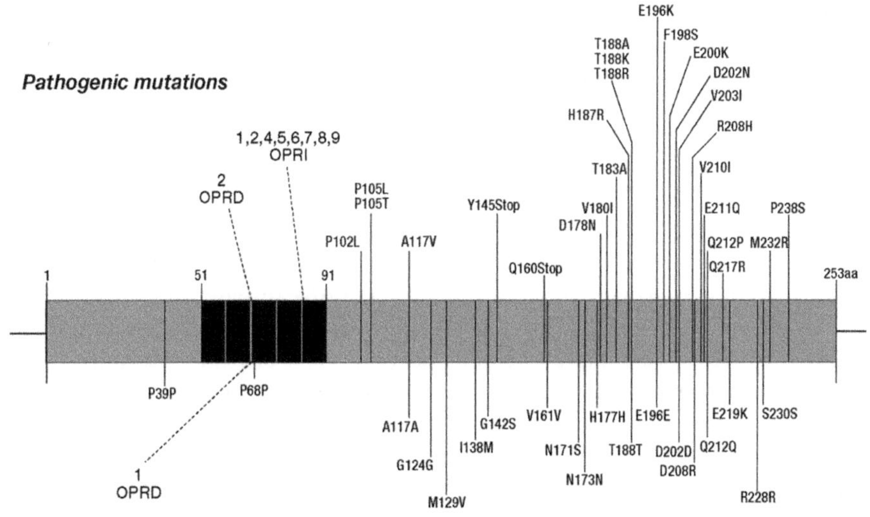

Figure 6.1 Schematic representation of coding changes in human prion protein. (Reproduced from Wadsworth *et al.*[36] with permission from BMJ).

The most common natural variation yielding polymorphism in the population occurs at position 129, encoding either a methionine or valine residue. Homozygotes (genotypes with two identical allele of the particular gene) are found to have a shorter disease incubation time when compared to heterozygotic individuals, both for transmissible forms of the disease like kuru[39] and innate entities like pathology associated with six-octapeptide repeat insertion.[40,41] In addition, this particular polymorphism determines disease susceptibility and predisposes individuals to diseases of sporadic and iatrogenic origin.[42,43] However, in inherited prion diseases it acts in *cis* with the mutation (*i.e.* both of them are on the same chromosome of homologous pair), while its role in the normal allele is insignificant, if any. Therefore, accurate identification of each *PRNP* genotype associated with genetic prion disease should take into account both mutation and also the amino acid present at codon 129 in the mutant allele (haplotype).[38]

Most *PRNP* gene mutations associated with familial CJD are located within the globular domain of the prion protein. Six of them (R148H, D178N, E196K, E200K, R208H, E211Q) could possibly change the salt bridge or hydrogen bond interactions and consequently the thermodynamic stability of the protein. However, only D178N and T183A yielded mutant PrPs with significantly altered conformational stabilities. In addition, the effect of the D178N mutation depends on the M129V polymorphism. The other factor distorted by pathogenic mutations of the gene encoding prion protein (D178N, T183A, E200K, OPR insertions) is the intracellular trafficking of the altered protein due to its partial misfolding. As a result, mutated PrP undergoes intracellular retention inside either the endoplasmic reticulum or Golgi apparatus and becomes a substrate for systems of proteasomal[44–46] or lysosomal protein degradation.[47–50] Other changes following the D178N point mutation disturb the ratio of the three major "glycoforms" of the prion protein. Asn181 and Asn197 are potential binding sites for N-glycans, and so we can have either the fully glycosylated form with two glycans, intermediate forms with only one glycan and the non-glycosylated form.[51] The under-representation of the latter has been shown in cell model studies and membrane fractions of FFI brains.[52] Mice models carrying the D178N mutation linked to Met at codon 129 spontaneously developed transmissible prions.[53] This again indicates the pivotal role of M129V polymorphism. However, the influence of *PRNP* mutations on the Prp glycosylation pattern is not limited to D178N, which, by analogy with E200K, favours the diglycosylated form of PrP. The reverse impact has been observed for G114V, T183A and V180I, with the diglycosylated form under-represented. The third group of mutations, with no apparent effect on glycoform ratio (predominance of the monoglycosylated forms), includes G114A, R148H, T188K, R208H, V210I, M232R and the insertion mutations.

The key role of the codon 129 genotype is reflected in the physicochemical properties of PrP^{Sc} profiles. The particular forms may differ in the size of the peptide products after protease treatment, proteinase resistance and the previously described glycoform ratio.[38] The two "types" which differ in the size of the proteinases-resistant core (21 kDa for type 1 and 19 kDa for type 2) and the location of the primary cleavage site (residue 82 for type 1 and residue 97 for type 2) have been characterized *in vitro* and interrelated with codon 129 of the mutated allele polymorphism.[54–56] Most PrP mutations in a significant association with Met at position 129 yield type 1 PrP^{Sc}, while type 2 is determined by Val129. The opposite trend has been observed for D178N and T183A, which express type 2 PrP^{Sc} with M129 in the mutant allele and type 1 PrP^{Sc} with V129 in the mutant allele, respectively.[38]

There is a strong indication that genetic CJD basically mimics the phenotypic range of sporadic variants of the disease, with both physicochemical properties of PrP^{Sc} and codon 129 genotype as main determinants of the phenotype.[38] This molecular feature combined with the identification of clinical and pathological aspects gives a potent tool for the analysis and interpretation of prion diseases, which may lead to significant improvements in our understanding of these disorders. On the other hand, it is clear that mutations of *PRNP* by themselves are not sufficient and that other genetic and environmental factors are necessary to cause the disease.

The identification of genes that contribute to pathogenesis is of paramount importance for understanding prion diseases.[57] So far, although many PrP interacting proteins have been identified, only a few of them affected prion replication.[8,58] A huge series of candidate genes has been analyzed for their effect on prion pathogenesis, and a positive effect has been elucidated for APP, the Alzheimer's disease precursor protein. Although the background of the effect is not known, knockout of the APP gene delays the onset of prion disease. An analogous outcome has been found for proinflammatory cytokine IL-1, while the opposite was found for anti-inflammatory IL-10.[57] A truly extraordinary conclusion has been reached for the effect of chemokines involved in neuroinflammation and their receptors. Abolition of expression of chemokines CCL2 and CCL5 leads to prolongation of survival times,[59] while no effect was found for their respective receptors, CCR2 and CCR5. This may be due to the complicated network of interactions recruiting the chemokines (*vide infra*). Although SOD1 over-expression and hypothetical SOD activity of the prion itself was initially found to be protective and life prolonging, recent findings indicate that it might only be the case for fast-replicating laboratory prion strains, thus making less likely a role for SOD mutations in human prion diseases.

In conclusion, genetic factors may have an impact on prion disease susceptibility and progression. The *PRNP* mutations usually yield proteins of lower thermodynamic stability or dramatically altered cellular trafficking. The actual impact of single mutations becomes evident in a way that is critically interconnected with genotype polymorphism, especially polymorphism of codon 129. Clearly, conversion of PrP^{C} to PrP^{Sc} involves a complex pathway that becomes even more intricate when correlation with other gene expression levels is taken into account.

6.3 Proteins Possibly Involved in the Disease

6.3.1 Scrapie Prion Protein

It's well accepted that the key event associated with prion diseases is the conversion of the normal cellular prion protein isoform (PrP^C) into the abnormal scrapie prion protein isoform (PrP^{Sc}). Prion is the infectious agent causing prion diseases[60] and it is mainly constituted by PrP^{Sc}, the misfolded form of PrP^C. PrP^{Sc} has a primary sequence identical to PrP^C but it exhibits physicochemical properties that differentiate it from the cellular isoform. Whereas PrP^C is an easily degradable monomer, PrP^{Sc} is aggregated and is highly resistant to protease digestion.[61,62] The protease-resistant core of PrP^{Sc}, termed PrP27–30, has a molecular mass of 27–30 kDa and is generated by the cleavage of the N-terminal 23–89 region.[7] Compared to PrP^C, PrP^{Sc} and PrP27–30 have a higher amount of β-structure and a lower α-helical content. It is known that prion fibres are characterized by a cross-β structure, but unfortunately the exact structural features of PrP^{Sc} have not been determined so far. Conventional solution NMR and crystallography have failed to solve the PrP^{Sc} structure because of its aggregated state and low solubility. A structural model of PrP27–30, obtained by means of electron diffraction, shows left-handed β-helices within region 89–175 associating in turn to form trimers (Figure 6.2).[63] The model was constructed by supposing that the

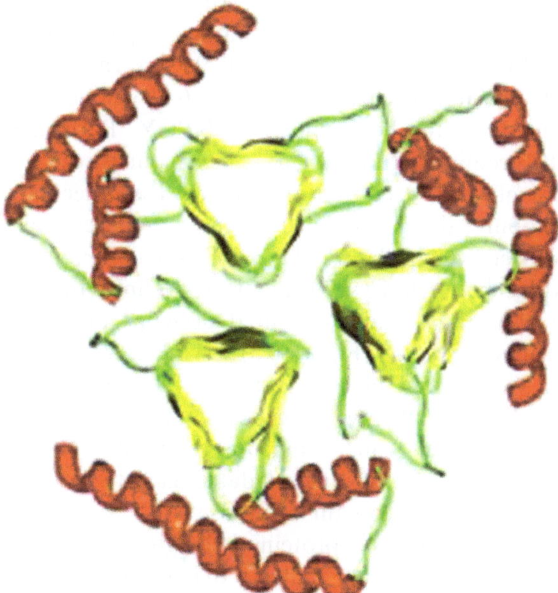

Figure 6.2 Trimeric model of PrP27–30.
(Reproduced from Govaerts *et al.*[63] with permission from the National Academy of Sciences).

remaining C-terminal sequence adopts its native α-helical structure. On the other hand, hydrogen/deuterium backbone amide exchange investigations and electron paramagnetic resonance data performed on recombinant PrP fibrils indicate that the fibrillar core is formed by residues 169–221 containing parallel β-sheets.[64,65]

6.3.2 Cellular Prion Protein PrP^C

The cellular prion protein (PrP^C) plays a key role in prion replication as well as in prion-induced neurodegeneration. PrP^C is a cell surface glycosylphosphatidylinositol (GPI)-linked glycoprotein enriched in detergent-resistant membranes (Figure 6.3).[66] PrP^C is encoded by the *PRNP* gene and is highly expressed in the central nervous system (CNS). PrP^C is synthesized in the endoplasmic reticulum (ER) as a 253 amino acid long protein. In the mature protein, the N-terminal signal peptide (residues 1–23) is cleaved after translation and a GPI anchor is attached at Ser230. PrP^C has two N-linked glycosylation sites at residues Asn181 and Asn197.[67] Some PrP^C undergoes proteolytical cleavage by cellular proteases near residue 111 to generate N- and C-terminal fragments called N1 and C1, respectively.[68,69] PrP^C is highly conserved amongst mammals, with amino acid identity of around 90%[70] and it has been also identified in birds, fish and *Xenopus laevis*.[70–72]

PrP^C is characterized by two distinct domains showing different structural and dynamic features.[73,74] The N-terminus (up to residue ~120) consists of a flexible and disordered region usually containing repetitions of specific amino acid sequences. In mammals, the region 60–91 is formed by the octapeptide PHGGGWGQ, repeated four times (Figure 6.3). In birds and reptiles the repeat unit consists of the hexapeptide PHNPGY. Both mammal and bird repeats have the peculiarity to bind Cu^{2+}, in a manner which is dependent on pH and metal availability (see Section 6.6). On the other hand, the PrP^C C-terminus is a globular domain formed by three α-helices and a two-stranded antiparallel β-sheet which flanks the first α-helix (Figures 6.3 and 6.4). A disulfide bridge between Cys179 and Cys214 links the second and third α-helices. The N-terminal and C-terminal domains are linked by a highly conserved hydrophobic region (Figure 6.3) serving as a transmembrane anchor rich in alanine, valine and glycine residues.[75] In mammals, this region is strongly associated with PrP amyloidogenesis and neurotoxicity and it contains a His residue acting as Cu^{2+} anchoring site (see Section 6.6).

The first structure of a prion protein dates back to 1996, when the NMR structure of the C-terminal domain of mouse PrP^C was solved.[76] Since then, many NMR structures of PrP^C proteins belonging to different species have been reported.[74,77,78] All of them indicate that the folding of the C-terminal domain is highly conserved among different species, supporting the well-established fact that the three-dimensional structure is retained through evolution much more than sequence similarity. All of the available PrP^C structures were obtained using soluble recombinant PrP, which might have structural rearrangements different from those of native PrP^C, a glycoprotein

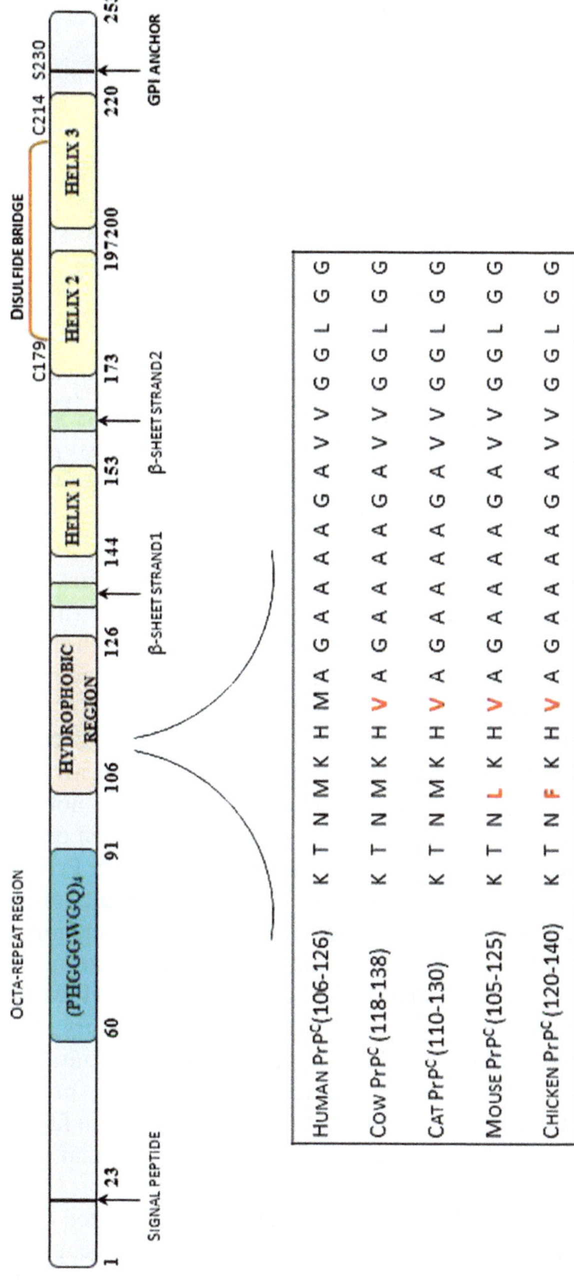

Figure 6.3 Representation of the cellular prion protein; the amino acid sequence corresponding to the hydrophobic region of PrP^C belonging to different species is shown in the box.

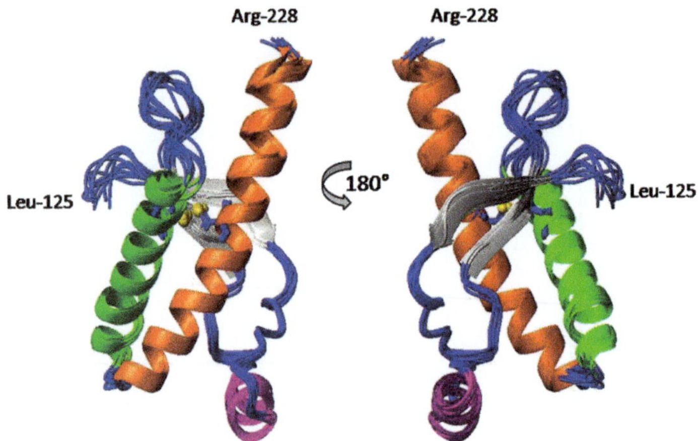

Figure 6.4 Visualization of the first 10 NMR structures of the human PrP^C globular domain (region 125–228) extracted from Zahn *et al.*[74] (pdb 1QLZ; RCSB protein data bank). Helix-1 (144–153), helix-2 (173–197) and helix-3 (200–227) are shown in *magenta*, *green* and *orange*, respectively. The two-stranded antiparallel β-sheets are shown in *grey*. The sulfur atoms of Cys179 and Cys214 are shown as *yellow spheres*.

anchored in a membrane. However, the fact that (i) native bovine PrP^C has a CD spectrum identical to that of the recombinant PrP[79] and (ii) the folding of recombinant PrP^C remains unchanged after membrane anchoring,[80] suggests that *in vitro* PrP^C adopts substantially the same three-dimensional structure as its recombinant form.

Although the first evidence identifying PrP^C as a normal cellular protein appeared in 1985[81] and since then an impressive number of investigations has been carried out, the physiological function of PrP^C is still not clearly established.[82] Most studies aimed at addressing this issue have been performed using genetically engineered mice in which the gene encoding PrP is disrupted (PrP-knockout mice).[83,84] Unfortunately, none of the results obtained so far has allowed unequivocal identification of the biological function of PrP^C. In fact the majority of the PrP-knockout mice studied have a normal lifespan and show no developmental or anatomical abnormalities.[83] However, in some cases, PrP-knockout mice showed alterations in olfactory function[85] or myelination[86] and exhibited slight changes at the behavioural and cellular levels.[87] No clues to PrP^C biological function have been provided by the identification of the structural features of recombinant PrP; in fact the PrP fold is relatively unique, with the only structural similarity to that of the Doppel (Dpl) protein,[88] despite their distinct origins[89] (*vide infra*).

Although no precise biological function has been attributed to PrP^C, many different biological functions have been proposed. One of them suggests a role for PrP^C in the regulation of ion channels and neuronal excitability. PrP^C is indeed localized at synaptic sites, and is important for normal synaptic development and function.[90] Cerebellar or hippocampal neurons derived from PrP-knockout mice exhibit a number of electrophysiological abnormalities.[91–93] PrP^C itself might form ion channels in the cell membrane, which might be responsible for PrP-related neurodegeneration. In particular, the expression of a PrP^C protein lacking the hydrophobic region (Δ105–125PrP) induced spontaneous ionic currents in transfected cell lines.[94]

Recent findings have also indicated that PrP^C may behave as a cellular adhesion molecule, participating in neurite outgrowth, neuronal survival and neuronal differentiation.[95] Moreover, it has been shown that PrP^C interacts with several surface proteins, such as laminin,[96] the 37-kDa laminin receptor precursor (LRP),[97] the 37-kDa/67-kDa laminin receptor,[98] the neural cell adhesion molecule (NCAM)[99] and the contactin-associated protein (Caspr), a neurite outgrowth-inhibitory factor.[100] The involvement of PrP^C in cell adhesion is also supported by recent studies performed in the zebrafish *Danio rerio*.[101]

Neuroprotective and pro-survival functions have been attributed to PrP^C as well. PrP^C overexpression is able to protect cell lines and primary neurons from apoptotic stimuli[102,103] and the N-terminal region of PrP can modulate changes in intracellular reactive oxygen species (ROS) caused by cellular stress.[104] The PrP^C complex with the cytoplasmic co-chaperone molecule, the stress-inducible protein 1 (STI1), promotes neuronal survival and differentiation;[105] moreover, the PrP^C–STI1 complex is also implicated in the self-renewal of neural progenitor and/or stem cells.[106] Collectively, these data suggest a role for PrP^C in protection from cellular stress, as well as neuronal survival, differentiation and proliferation. On the other hand, the deletions and point mutations in the protein can be responsible for cytotoxicity in cultured cells and neurotoxicity in mice. The evidence of both neuroprotective and neurotoxic properties of PrP^C suggest that prion toxicity may be due to suppression of the physiological function of PrP^C.[107]

6.3.3 Prion-like Proteins

PRNP, *PRND* and *SPRN* are the three members of the prion gene family encoding PrP^C, and its two paralogues, the Doppel (Dpl) protein and Shadoo (Sho) protein. Functional interactions between these proteins are capable of modulating the degeneration of cerebellar cells in culture or in transgenic mice, with Doppel having opposite functions to either PrP^C or Shadoo.[108]

Dpl is a PrP-like protein which has a critical role in the proper functioning of the male reproductive system.[109] Dpl shows neurotoxic properties in the absence of PrP^C and causes degeneration of Purkinje neurons in the cerebellum of mice.[110] Dpl, a 125 amino acid residue protein, is thought to be an N-truncated form of PrP^C, sharing several structural and biochemical features with it. As previously mentioned, Dpl has a tertiary structure very similar to the C-terminal domain of PrP^C, despite the low ($\sim$25%) sequence identity. As with PrP^C, the N-terminal signal sequence of Dpl is cleaved and a GPI anchor is attached to its C-terminus. Dpl, like PrP^C, has two N-linked glycosylation sites at Asn99 and Asn111. Like PrP^C, Dpl localizes to lipid rafts. However, contrary to PrP^C, Dpl lacks the octarepeat binding domain and does not bind copper. These features make Dpl highly homologous to the C-terminal part of PrP^C.

Sho, a protein expressed in the CNS, has remarkable similarity to PrP^C.[111,112] Murine Shadoo is a 98 amino acid protein whose architecture is highly reminiscent of the PrP^C N-terminal domain. Sho lacks the octarepeat sequences present in PrP^C, but it possesses repetitions of tetrapeptides rich in glycine, serine, alanine and arginine. Sho, like PrP^C, possesses a hydrophobic domain highly homologous to those found in PrP sequences from diverse species, thus suggesting possible functional homology. Like the N-terminal domain of PrP^C, Sho is an almost unstructured protein, with an α-helical domain much shorter than those found in PrP^C and Dpl protein and with no disulfide bridges. Like PrP^C, Shadoo is a GPI-anchored membrane glyco-protein which undergoes endoproteolysis. Shadoo is mainly expressed in the Purkinje cells in the cerebellum and in the dendritic processes of hippocampal pyramidal cells. Since these two brain regions are characterized by the absence of PrP^C, it is believed that Shadoo could have a PrP^C-like function in those areas of the brain naturally deficient in PrP^C.

6.4   Pathology Associated with Proteins

Prion diseases, also known as transmissible spongiform encephalopathies (TSEs), are neurodegenerative disorders affecting both humans and animals. Prion diseases, similar to the neurodegenerative pathologies of other diseases such as Alzheimer's disease, Parkinson's disease and Huntington's disease, are characterized by progressive neuronal dysfunction leading to cell death and by the presence of protein-rich aggregates in the brain.

A number of different prion diseases affecting diverse mammal species are known, which may be sporadic, inherited or acquired by infection (Table 6.1). Kuru was the first human TSE transmitted to experimental animals and it is now widely accepted that it was transmitted among the Fore people of Papua New Guinea through cannibalism.[113] Among human prion diseases, familial

Table 6.1   TSEs: names, affected animals and ethiology.

Prion disease	Affected animals	Ethiology
Kuru	Human	Infection
Iatrogenic Creutzfeldt–Jakob disease (CJD)	Human	Infection
Sporadic Creutzfeldt–Jakob disease (CJD)	Human	Unknown
Familial Creutzfeldt–Jakob disease (fCJD)	Human	Genetic
Variant Creutzfeldt–Jakob disease (vCJD)	Human	Infection
Gerstmann–Sträussler–Scheinker (GSS) disease	Human	Genetic
Fatal familial insomnia (FFI)	Human	Genetic
Scrapie	Sheep and goats	Infection
Bovine spongiform encephalopathy (BSE)	Bovine	Infection
Chronic wasting disease (CWD)	Deer and elk	Infection
Feline spongiform encephalopathy (FSE)	Cats	Infection

CJD (fCJD), FFI and GSS diseases are caused by mutation of the PRNP gene.

TSEs are the only neurodegenerative diseases known to be infectious and able to be transmitted between individuals.[6] Individuals affected by prion disease show both cognitive and motor dysfunctions. In addition, the disease is characterized by the propagation of an infectious agent called a prion and by the presence of amyloid deposits called plaques. As described by the prion hypothesis,[1,60] the agent causing prion diseases is mainly composed of PrP^{Sc}, a misfolded form of the cellular prion protein (PrP^{C}), termed "prion" by Prusiner 30 years ago.[114]

The transmissibility of TSEs is regulated by species barrier and infectivity between heterologous species is very unlikely. Probably, the main element controlling the interspecies transmission of prion disease is the similarity of the PrP primary structure between different species. In particular, the strong species barrier between mice and hamster might be explained by considering that mice and hamster PrPs have distinctly different amino acid sequences.

Compared to other known infectious pathogens, prions are devoid of nucleic acid and they consist exclusively of a protein (PrP^{Sc}). More than 10 years ago, protein misfolding cycling amplification (PMCA) allowed the replication of prions *in vitro*.[115] The prions which were obtained in this way have very similar biological, biochemical and structural properties to prions produced *in vivo*, strongly supporting the prion hypothesis.

In all TSEs, the key pathological event is the conformational conversion of PrP^{C} into PrP^{Sc}, which abnormally replicates itself to generate amyloid structures. The exact mechanism of conversion between the two PrP isoforms is still unknown, but it seems that PrP^{C} and PrP^{Sc} associate in an intermediate complex during the formation of the pathological isoform.[116] Although many results indicate that PrP^{C} expression is necessary to initiate the disease, it is still not clear if PrP^{C} acts as a substrate or as precursor for PrP^{Sc} formation.[117] Moreover, it has been shown that the GPI membrane anchor is necessary for disease development. Mice expressing anchorless PrP show negligible clinical symptoms when infected with PrP^{Sc}.[118] Two different models have been proposed to explain the PrP^{C}/PrP^{Sc} conformational conversion (Figure 6.5):[6]

1. The template-direct refolding model proposes that endogenous PrP^{C} might interact with PrP^{Sc} (the infectious agent), inducing the transformation of PrP^{C} into further PrP^{Sc}. It is hypothesized that the PrP^{C}–PrP^{Sc} interaction might favour conversion of PrP^{C} into PrP^{Sc}, a process normally non-spontaneous and characterized by a high energy barrier.
2. The seeded nucleation model postulates that PrP^{C} and PrP^{Sc} conformations are in thermodynamic equilibrium, normally shifted well towards the cellular isoform. It is thought that the infectious agent is an oligomer of PrP^{Sc} molecules. The infectious oligomer acts as a seed to bind further monomeric PrP^{Sc} to form amyloid. New infectious seeds can be formed by the fragmentation of PrP^{Sc} aggregates (amyloid), resulting in replication of the infectious agent.

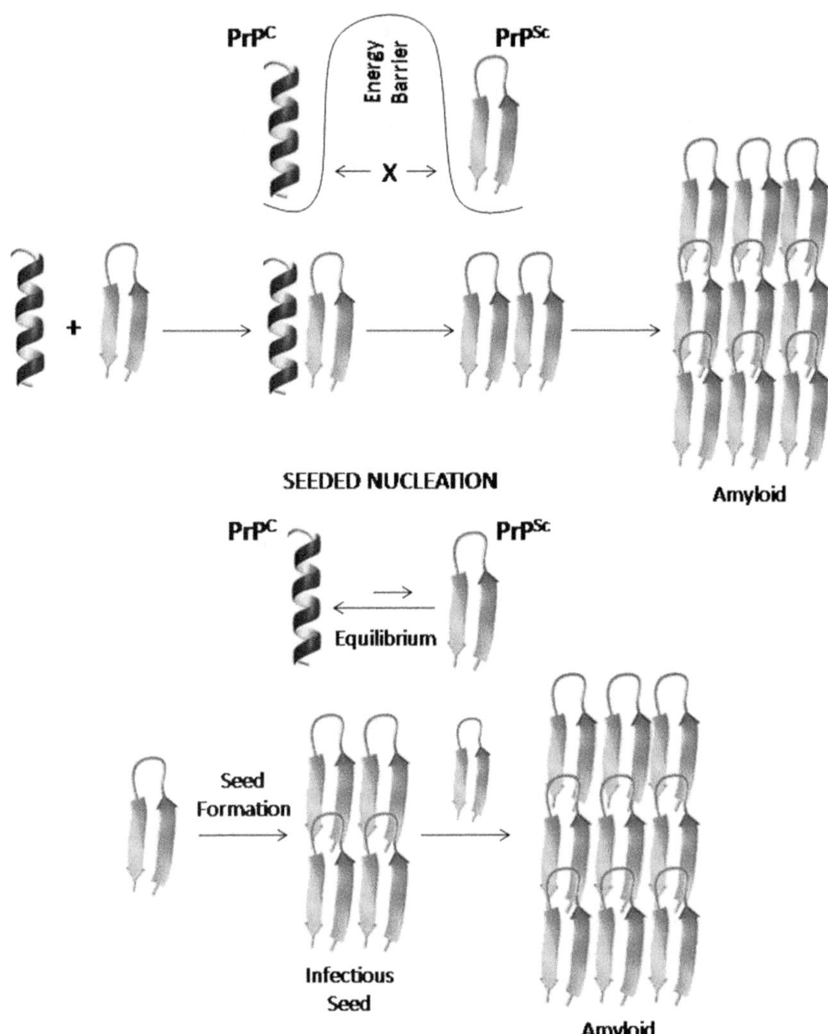

Figure 6.5 Models proposed for PrP^C conversion to PrP^{Sc}. (Adapted from Aguzzi and Calella[6]).

Several lines of evidence indicate that, besides PrP^{Sc}, other co-factors might be involved in prion replication.[119] Synthetic RNA might behave as a conversion factor, accelerating prion formation in hamsters.[120] RNA might participate as a scaffold for PrP^C conversion on which electrostatic interactions between the negatively charged RNA and PrP might occur.[121] Moreover, the additional presence of synthetic phospholipids[122] also enhances PrP^{Sc} replication, supporting the higher infectivity detected with lipid membrane-associated PrP^{Sc}.[123]

Investigations on inherited human prion diseases, including GSS, FFI and fCJD, have provided useful hints to better understand the mechanism of prion replication. These prion diseases are associated with mutations in the *PRNP* gene which yield specific point mutations in the amino acidic sequence of PrP[C] able to induce spontaneous formation of PrP[Sc] in the brain. It is believed that inherited point mutations might thermodynamically destabilize PrP[C], thus (i) favouring the misfolding process,[124,125] (ii) triggering abnormal interactions with other cofactors[126,127] or (iii) promoting aberrant prion accumulation inside the cell.[128] The Met/Val129 polymorphism affects the familial prion disease phenotype and among the inherited point mutations D178N, E200K and Q212P correspond to FFI, fCDJ and GSS, respectively. NMR and molecular dynamics investigations performed on mutated human PrP proteins could correlate these point mutations with specific alterations in the PrP structure. In particular, for both E200K and Q212P mutations a larger flexibility of the loop between helix-2 and helix-3 and the loop between β-sheet-2 and helix-2 is observed in both PrP mutants regardless of the position mutation.[129]

It is also believed that PrP[Sc] obtained from different strains has different conformational or aggregation states. All TSEs are caused by different strains characterized by distinct (i) incubation times, (ii) patterns of protein aggregate deposition, (iii) clinical features and (iv) specific neuronal target areas. New prion strains usually appear upon prion transmission across an interspecies barrier or in animals of the same species expressing different polymorphisms of the prion gene.

As previously mentioned, amyloid plaques constituted by PrP are typically found in the brains of human and animals affected by prion diseases. However, it is believed that the PrP[Sc] oligomers, rather than amyloids, may constitute the nucleation centre leading to prion replication.[130,131] In order to correlate prion infectivity with the aggregate size, the effects of different fractions with different molecular masses, derived from the fragmentation of PrP[Sc] aggregates, were evaluated in hamster.[21] These data showed that the most infective particles correspond to PrP oligomers containing between 14 and 28 PrP molecules, suggesting that lower molecular weight species are the main toxic particles. As with other neurodegenerative diseases, such as Alzheimer's disease, the formation of plaques may therefore be a protective consequence of over-production or mishandling of misfolded and abnormal proteins rather than a primary cause leading to the disease.

Prion amyloids exhibit features very similar to those of other amyloids. They were first identified as rod-shaped particles in fractions enriched for scrapie infectivity[132,133] and they appear to be identical to other amyloids. Analogously to Aβ fibrils, prion rods can be detected by means of spectroscopic measurements, using small dyes such as Congo red, thioflavins and their derivatives. These molecules can selectively interact with the cross β-sheet conformation of protein aggregates, giving rise to enhanced fluorescence (thioflavins) or to a red-shift of the UV–visible absorption spectrum (Congo red). Unfortunately these methods are not able to recognize oligomeric, prefibrillar species and amyloid species with no β-sheet core. Prion-like amyloids are present in lower eukaryotes as well. Fungal prions are molecules containing proteins different from PrP, like HET-s, Ure2p and Sup35. These proteins, like PrP, can adopt both non-amyloid and amyloid structures, but contrary to PrP, their conformational conversion is associated with important physiological functions in yeast, such as regulation of gene transcription and translation and self-recognition.

6.5 Neuroinflammation

Even though inflammation is a well-recognized phenomenon and well-researched area, the precise definition of inflammation remains hazy.[134] It can be defined in clinical, pathological and molecular terms. Clinically, prion disease patients do not show any of the classical inflammatory symptoms of Celsus,[a] namely *calor* (warmth), *dolor* (pain), *tumor* (swelling) and *rubor* (redness). Evaluation of inflammation at the histopathological level reveals that, although there are no signs of the presence of neutrophils, cells usually associated with typical acute response, the recruitment and activation of brain macrophages (microglia)[135] and perhaps astrocytes (astroglia) occurs. For that reason it is often termed atypical brain inflammation.[136] The lack of acute response (usually associated with the release of histamine and resulting vascular reactions leading to tissue destruction) helps to maintain the post-mitotic population of neurons and efficient neuronal transmission. For that reason, the presence of microglia activation could be considered the neurological form of chronic inflammation, at least at its early phase.[137]

The glial activation could result from a direct stimulatory effect of either diffuse or plaque PrP^{Sc} deposits and occurs early during prion infection.[135] Alternatively, it might be the consequence of the prior activation of T cells by the PrP^{Sc} deposits in the peripheral tissues, followed by their migration through the blood–brain barrier (BBB) and release of cytokines that activate brain macrophages.[138] Activated glial cells stimulate the expression of proinflammatory proteins, including complement factors, MHCII complex, interleukins, cytokines and chemokines.[139] Subsequent progression depends on the level of these factors and the time course of activation. Moderate conditions would enhance antigen presentation, thus promoting phagocytosis of stimulatory factors, and as a consequence the neuroprotective activity of microglia.[140] Although no direct evidence supports the pivotal role of these cells in the clearance of PrP^{Sc} deposits in the brain, the active participation of spleen macrophages in the process has been reported, thus supporting the hypothesis.[141] The worst case scenario corresponding to excessive or persistent stimulation of glial cells would lead to overstimulation of the immune system, which generates oxidative stress and in consequence promotes neurotoxicity, resulting in neurodegeneration.[4]

---

[a]Aulus Cornelius Celsus is the Roman author of an early encyclopedia of medicine, in which he is credited with recording the cardinal signs of inflammation.

6.5.1 Cytokines

The preliminary step that might be of critical importance for the neurotoxic activity of microglia is associated with the release of IL-1, the key proin-flammatory cytokine. The consequences of its activity are related to the level of its expression, which may be modulated by the major anti-inflammatory cytokine IL-10.[142] The low level of PrP^{Sc} accumulated in the brain could support the regulatory action of IL-10 and enable a moderate neuroprotective response of the immunological system. When the accumulation of PrP^{Sc} exceeds the critical concentration, augmented IL-1 mediates the subsequent activation of astrocytes through chemokines targeting CXC ligand receptors. Astrocyte activation starts and accelerates the "snowball" mechanism, leading to overactivation of the immune system with subsequent production of oxidative stress and lethal neurotoxicity.

6.5.2 Chemokines

Elevated chemokine expression levels have been observed in numerous pathologies of the brain, ranging from viral and bacterial infections to multiple sclerosis,[143–145] stroke[146] and Alzheimer's disease,[147,148] indicating their important role in acute and chronic neurodegeneration.[149,150] Moreover, overexpression of chemokines is known to be one of the hallmarks of reactive gliosis leading to neurodegeneration.[151–153] The precise contribution of chemokines in disease progression and prion pathogenesis is unknown for the reason that most of them may target more than one receptor and most receptors bind more than one ligand, thus forming a complicated network regulated at multiple levels. The central crossing point of this network relies on chemokine receptor CXCR3 interactions. This receptor is widely expressed in astrocytes, microglia and neurons. Its widespread expression makes this protein an important interface in the signal transduction between glial cells and between glial cells and neurons. Depending on the progress of the disease, the effects of activation of these signals might be either positive at the asymp-tomatic stage, when increased microglia tropism leads to more effective clearance of prion deposits, or negative at the symptomatic stage of disease. Activated astrocytes and neurons may stimulate microglial cells by the chemokine-driven activation of the CCR5 receptor, thus enhancing the excessive response of the immunological system.[154] Another critical chemokine, CCL2, known as monocyte chemo-attractant protein-1, is overexpressed in prion infected brain.[59,155] Its high affinity receptor, CCR2, is involved in acti-vation of microglia and recruitment of peripheral macrophages into the brain.[156]

6.5.3 Complement Cascade

Extracellular PrP deposits, either directly or through the activation of microglia, initiate the complement cascade which usually provides an innate immune mechanism against pathogenic microorganisms. In the CNS the proteins of the complement system are expressed by microglia, astrocytes and even neurons.[157] The typical activation complement cascade involves a series of protease-like protein complex assemblies which generate the major component of the membrane attack complex (MAC) facilitating cell lysis.[158] The complement may have dual roles, both protective and detrimental in neuro-degeneration and neuroinflammation. The PrP amyloid plaques are immuno-labelled for complement components but not for immunoglobulins.[159] Production of complement proteins is a source of proinflammatory peptides that induce chemotaxis of microglia. When the level of activation is moderate, it may facilitate the clearance of the PrP deposits by microglia, thus contributing to the neuroprotective role of the complement. However, subsequent activation of the complement cascade may facilitate the transition of the system to the stage where detrimental conditions dominate. Active compounds of the complement system, such as C1q (complement component 1q) and C3b (the product of hydrolytic cleavage of complement component 3), are present in disease-associated PrP deposits that coincide with MAC in affected neurons.[160] The former serves as a multivalent recognition molecule of primary component complex C1, which activates the classical pathway of the complement cascade. It is known to bind to damaged neurons and amyloid deposits.[161] The amyloid deposits activate complement component 5 to yield the product involved in MAC formation.[158]

To cut a long story short, in prion diseases the immunological response could be both protective and detrimental, depending on the level at which prion protein deposits stimulate the expression and synthesis of proinflammatory factors and the severity of immunological response.[137] Inflammation in the central nervous system is thus a double-edged sword.

6.6 Metals Involved; Pathways Activated by Metals

6.6.1 Metal Binding Sites in PrP

As discussed in the previous section, PrP^C possesses a flexible N-terminal region (residues 23–120) and a conserved globular C-terminal domain (residues 121–231) (Figure 6.3). The N-terminal part of the human PrP^C consists of four PHGGGWGQ octarepeats (residues 60–91), flanked by two positively charged clusters (residues 23–27 and 95–110). Histidine imidazoles from the octarepeat regions (His61, His69, His77 and His85), and also His96 and His111 from the neurotoxic part, are potential metal binding sites.[162] Although the physiological role of PrP^C is poorly understood, it is well accepted that PrP^C is a copper binding protein. A number of studies has been performed in order to clarify the cellular PrP^C function associated with copper binding.[163] One of the major breakthroughs of these studies was the finding that Cu^{2+} and Zn^{2+} stimulate the endocytosis of PrP^C, thus linking metal binding to a physiological response.[164,165]

Since one of the major postulated roles of PrP^C is the sequestration of Cu^{2+}, we will concentrate on the coordination mode and binding parameters of this metal in the next subsection. Later, the binding of other metals ions, such as Zn^{2+}, Fe^{2+} and Mn^{2+}, will be discussed. These metal ions are also able to bind to PrP, although with much lower affinity.[166] Among the investigated metals, the coordination of Zn^{2+} has been most thoroughly assessed; it may also be a natural PrP^C ligand *in vivo*, due to its high synaptic concentration. Some groups reported Zn^{2+} binding *via* one or two octarepeat histidine side chains, with saturation levels achieved after one Zn^{2+} equivalent is bound;[167] we will discuss this question and the possible competition with copper ions in binding to PrP in the following subsection. Then, we will briefly summarize the effect of the metal binding on the folding and fate of the prion protein.

6.6.1.1 Cu^{2+}

Copper has been the major focus of the interaction of metals with prion proteins, since its binding both to the full-length protein and to its fragments[26,163,168] is thermodynamically preferred over other metal ions. At physiological pH, naturally occurring PrP^C can bind between one and five or six Cu^{2+} ions *in vivo* while retaining its solubility.[169] The primary coordination site for copper is the octarepeat domain, and the binding of the metal promotes structuring of the otherwise unstructured N-terminal part.[170] Each of the four octarepeats (–PHGGGWGQ–) can bind one copper ion, which means that up to four copper ions can bind to the whole octarepeat region.[171] The first available Cu^{2+} binds exclusively to three or four His imidazole side chains with a K_d of about 0.1 nM.[172] After saturation is achieved, Cu^{2+} can bind two additional histidines outside the repeat region.[162] However, the difference in affinity of the so-called octarepeat and the neurotoxic regions is still under debate and has recently been reviewed.[173,174]

Before we go into detail, and take a closer look at the binding of metal ions to different families of PrPs, we must mention that the although the similarity of these proteins is not remarkably high between different classes,[175,176] the domains crucial for both structure and metal binding are well-conserved between various animal species.[177–179] In the avian variant of PrP^C, the repeat region (residues 53–94) containing the hexapeptide unit (–PHNPGY–) binds Cu^{2+} in a way highly comparable to the human variant.[180,181] The avian sequence itself is interesting from a chemical point of view, since the proline residue in the middle might act as a coordination break point.[182]

Piscine prion proteins have repeat units capable of binding copper, and the repeat domains of PrP-like proteins (zPrP) from *Takifugu rubripes* (pufferfish) and *Danio rerio* (zebrafish) (residues 94–128 and 60–87, respectively) are formed by sequences that contain two histidine residues (–GHGYGVYGH– for *T. rubripes* and –HXGHXG– for *D. rerio*).[173] What is of particular interest is the fact that copper binds to these fragments with higher affinity than it does to the human octarepeat domain.[183,184] The reason for this might be because of the presence of an additional His in the fragment: the Cu^{2+} complex formed with the *T. rubripes* St2PrP96–107 peptide fragment at pH 7.4 has a {$2N_{Im}$, $2N^-$} set of donor atoms, and the presence of two imidazoles makes Cu^{2+} binding much more effective than its binding to the single human octarepeat unit.

The hydrophobic domain between the tandem repeats and the C-terminal globular region is strongly associated with PrP amyloidogenesis, neurotoxicity and proteinase K resistance.[185] Although this domain is highly conserved between different vertebrate PrPs, only humans and several other mammals (such as cattle, deer, elk, sheep, goats, mink, cat and antelopes) are affected by prion diseases. There might be numerous reasons for this; the lifespan of the species may also need to be taken into account.

For several years, it has been a priority for numerous research groups to understand the differences in dynamics, folding and metal binding affinities of PrPs from different species. Attempts were made to understand the specific copper binding affinity to human (h), mouse (m), chicken (ch) and piscine PrP using model peptides. This approach gives detailed thermodynamic information about the binding of Cu^{2+} to the unstructured regions of PrPs and allows some understanding of the difference in the binding between different species from a chemical point of view. In what follows, we try to summarize the most recent findings about the interactions of copper with single repeat units, multi-repeat units and with the amyloidogenic region, which contains two additional His residues, which are potential copper binding sites. What is clear is that the single repeat units from the species mentioned above are able to coordinate a single Cu^{2+} ion.[182,186–192]

The His imidazole ring acts as a copper anchoring site; it is the primary binding site for the metal ion. At slightly higher pH, the main chain amide nitrogens begin to participate in the coordination, forming a square-planar copper complex. Taking the human PrP as an example, the Cu-hPrP60–67 complex exists in its $CuH_{-2}L$ form at physiological pH, with a {N_{im}, $2N^-$, CO} metal coordination sphere and a distorted planar geometry. In this case, Cu^{2+} is bound to the imidazole side chain of His61, to two amide nitrogens from residues Gly62 and Gly63 and to the carbonyl oxygen of Gly63 (Figure 6.6).[193] Most probably, the coordination sphere is completed by a water molecule in the apical position. Trp99 from the analyzed region folds towards the central Cu^{2+} ion, most probably affording a hydrophobic protection that stabilizes the complex.[187] It might also have a role in the reduction of Cu^{2+} to Cu^+.[194,195]

Once the single repeat coordination mode has been established, the binding of Cu^{2+} to the multi-repeat region can be discussed. Two different binding modes have been characterized, depending on the pH and the concentration of copper ions. In the first proposed binding mode, the so-called inter-repeat binding, only the His imidazoles are the binding sites for the Cu^{2+} ion, resulting in a single copper ion bound to four His imidazoles (Figure 6.7A); this binding mode occurs at equimolar copper concentrations, in the pH range 5.5–7.5.[196] The second type, the intra-repeat binding, occurs at physiological pH (higher than the pH of the "inter-repeat binding") and at higher Cu^{2+} concentrations. Its binding mode is exactly the same as the one found in the single repeat

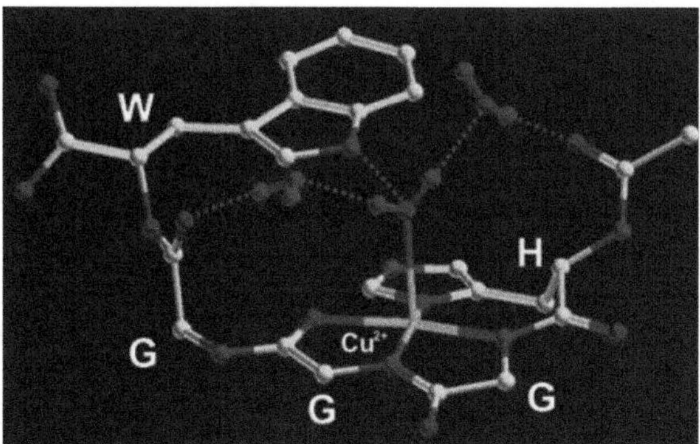

Figure 6.6 Crystal structure of the copper complex with the HGGGW sequence of the hPrP single repeat unit.
(Adapted from Burns *et al.*[193]).

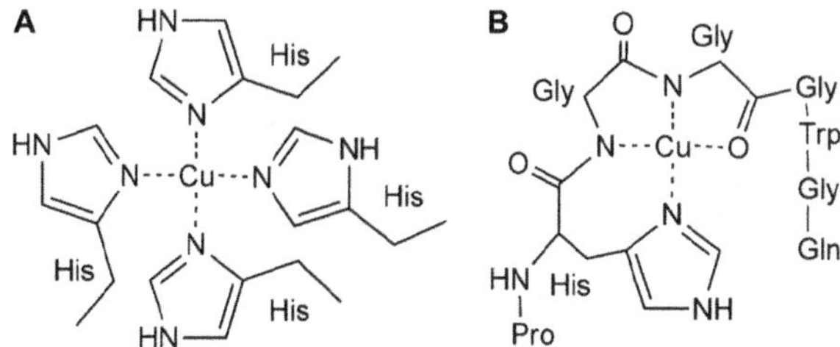

Figure 6.7 Representation of (A) inter-repeat and (B) intra-repeat copper binding modes within the hPrP octarepeat region.

unit: distorted square planar, with the $\{N_{im}, 2N^-, CO\}$ donor set (Figure 6.7B).[197] The number of bound Cu^{2+} ions is equal to the number of repeat units (four in the case of hPrP), with the His imidazole from a repeat region being the anchoring site for copper. Both of these binding modes, found by the short peptide model approach, have also been identified in the full-length hPrP.[198] The idea that the number and coordination mode of Cu^{2+} is crucial for the structure of the whole N-terminal part of PrP, and the fact that both of these features are affected by slight changes in pH, may explain the potential involvement of human prion protein in copper trafficking *in vivo*.[199,200] The same role is proposed for prion proteins from other species, since the intra-repeat and inter-repeat binding modes have also been well documented in the tandem repeat regions of chicken and piscine PrP.[183,188]

As far as the binding of copper outside the octarepeat region is concerned, two other histidine residues, placed directly outside the repeat region (His96 and His111 in the case of hPrP), are involved in Cu^{2+} binding and might play a role in the structural rearrangements that lead to the formation of the pathogenic isoform. The hydrophobic prion region, hPrP91–126, which involves both of the mentioned histidines, has been intensively studied in the past few years; different metal binding preferences have been proposed and the question of the stabilities and structures of the two independent Cu^{2+} binding sites seems to be still a matter of debate.[162,201–203]

In principle, there are contradictory data on the number of copper ions that can be coordinated outside the octarepeat region. Klewpatinond *et al.* used CD spectroscopy to show that one copper can bind to His96 and His111 in a multi-His mode,[204] while NMR and potentiometric analysis performed by Remelli *et al.* support the existence of two independent binding sites (located respectively at His96 and His111), excluding the occurrence of multiple His binding modes.[205] Various studies on the hPrP91–114 region show that both His96 and His111 can bind Cu^{2+}, although His111 seems to be the preferred binding site.[205,206] However, no consensus has been reached on this issue, since there is also evidence indicating His96 as the higher affinity copper binding site.[207] To clarify this issue, the thermodynamic and structural characterizations of copper complexes with short model peptides containing each of the two histidyl residues were performed. Various studies prove that the His imidazole is the anchoring site for copper, and at higher pH this anchoring is followed by subsequent amide nitrogen deprotonation, resulting in a $\{N_{im}, 2N^-, O\}$ binding mode at pH 6.5 (where the oxygen is either an adjacent carbonyl group or a water molecule) and a $\{N_{im}, 3N^-\}$ coordination mode at around pH 8.5.

The thermodynamic stability studies of copper complexes with model peptides containing the crucial His96 and His111 as the first or the last residue in the sequence (hPrP91–96, hPrP96–100, hPrP106–111 and hPrP111–115) have recently been carried out.[208] The Cu^{2+} ion was expected to anchor at the His imidazole and successively deprotonate backbone amide nitrogen atoms, either towards the N-terminus for fragments having His at the C-terminus (hPrP91–96, hPrP106–111) or towards the C-terminus for peptides with His at the N-terminus (hPrP96–100, hPrP111–115). As far as His96 is concerned, copper amide deprotonation can occur either in the N-terminal or C-terminal directions, leading to the formation of six- or seven-membered chelate rings, respectively. The most interesting outcome of the study was that the thermodynamic parameters of both possible His96 coordination modes are very similar (although normally the N-terminal direction would have been expected, resulting in a thermodynamically favoured six-membered chelate ring). NMR measurements show that the involvement of Ser97 and Gln98 amide groups in the binding of copper brings the Trp99 indole close to the central metal ion. Such an orientation of the aromatic ring allows a possible Cu^{2+} ion–π interaction and might explain the unexpected stability of the seven-membered chelate ring.[209]

For the His111 site, as thermodynamically expected the amide deprotonation towards the N-terminus is favoured, although the formation of minor species with "C-terminal direction preferences" cannot be excluded. Moreover, it is not yet clear whether the Met thioether sulfur can participate in the coordination of Cu^{2+}.[210,211] Theoretical calculations suggest that the Met109 sulfur is 4.7 Å away from copper only for the $\{N_{im}, 2N^-, O\}$ species, present at pH 6.5. NMR studies show that for the 3N complex, this distance is in the range 4.1–5.0 Å, while a greater distance of 5.3 Å is found for the structure of the 4N complex

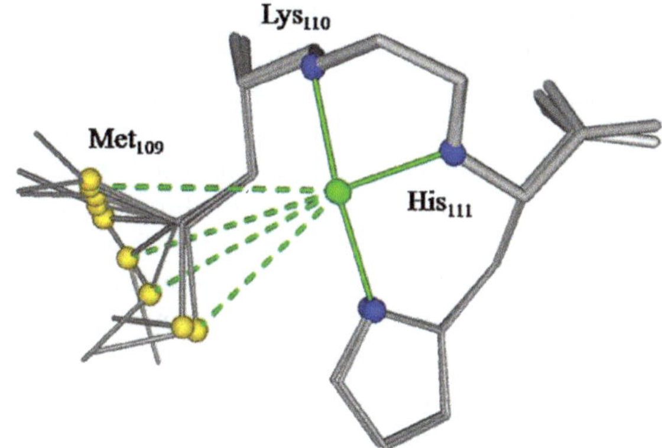

Figure 6.8 Selected NMR structure of a Cu^{2+}–hPrP106–111 complex showing the $\{N_{im}, 2N^-, O\}$ donor set.
(Adapted from Migliorini *et al.*[188]).

at pH 8.5 (obtained using Ni^{2+} as diamagnetic probe for Cu^{2+}).[212] The involvement of this Met residue (even taking into consideration simple far-distance interactions, Figure 6.8) could play a key role in stabilizing the His111 site, and it could very well explain the difference in affinities of the His111 and His96 regions towards Cu^{2+}. Recent thermodynamic studies prove that His111 is the preferred outer-octarepeat copper binding site (the K_d values are 160 and 58 nM for His96 and His111, respectively).[174]

Since the copper complexes of both His96 and His111 sites have the same donor sets ($\{N_{im}, 2N^-, O\}$ and $\{N_{im}, 3N^-\}$ at higher pH), the preference of copper for His111 must be a result of other long-range interactions with the side chains of neighbouring residues: the interaction with the thioether sulfur of Met109 seems to be a convincing explanation. It can further be supported by the example of the mouse prion protein: Met109 is well conserved in all mammals except mice. This variant lacks both Met109 and Met112, and in this case the binding of copper to His95, and not to His110, is preferred (they correspond to the human His96 and His111, respectively).[162,213]

ChPrP also contains two additional His sites: His110 and His124, which correspond to hPrP His96 and His111 in the human variant. They are good copper binders and their Cu^{2+} binding modes are the same as those of the human hydrophobic fragment (square planar geometry and a $\{N_{im}, 2N^-, O\}$ or $\{N_{im}, 3N^-\}$ donor set, depending on pH).[213] In this case, however, only the coordination towards the N-terminus was observed. From the thermodynamic point of view, the binding of copper to His110 (corresponding to hPrP His96) is preferred: the trend is reversed from the one observed on the same region in human prions. This inverted preference might result in a different conformational rearrangement of the hydrophobic region. The chicken prion protein, with preference for the "first" His110, does not fold into the pathogenic form, whereas in the human variant the copper preference for the "second" His111 might be related to the conversion of PrP^C to the proteinases-resistant isoform.[188] In addition to these specific binding sites, PrP^C can also bind Cu^{2+} non-specifically, when high concentrations of these ions are present. As a result, PrP^C can bind 10–11 Cu^{2+} ions at pH 4, when the octarepeat region shows no binding and the C-terminal can bind only four ions.[214,215]

As far as the thermodynamic parameters of Cu^{2+}–prion or Cu^{2+}–prion fragment complexes are concerned, a variety of different K_d values has been reported. The first constant found for the Cu^{2+}–hPrP octerepeat region (in a 4:1 stoichiometry) was 6.7 µM.[216] This value is similar to that of 5.9 µM, reported two years later for a longer peptide fragment (residues 23–98).[163] The K_d of 6 µM for a single Cu^{2+} complex with a two octarepeat peptide fragment (residues 51–75) is also in good agreement with the values mentioned above.[217] A study by Witthal *et al.* shows that at physiological pH the N-terminal domain of PrP can bind approximately five copper ions with K_d values ranging from micromolar to nanomolar.[218] Surprisingly, Jackson *et al.* reported that the N-terminal octarepeat region binds only a single Cu^{2+} ion with a K_d of 10^{-14} M,[219] while Garrnett *et al.*, in a study that made use of small amounts of Gly and His as PrP competitors for Cu^{2+} binding, claimed that the affinity is in the micromolar range.[220]

Numerous other studies tried to determine the prion (and prion fragments) affinity towards copper ions, but to date no consensus has been reached on this question, with K_d estimates ranging between 10^{-6} and 10^{-14} M. This surprising variability of eight orders of magnitude depends on the sequence studied and the experimental conditions, such as the choice of an appropriate buffer or the presence of additional ligands (reviewed recently[174,219,221]).

6.6.1.2 Zn^{2+}

The presence of six histidines (in the case of hPrP: His61, His69, His77, His85 from the octarepeat region and His96 and His111 from the neurotoxic region) in PrP strongly suggests that it will be able to bind Zn^{2+} as well. However, there have been few studies of zinc–PrP complexes.[166,222] The metal's affinity for the protein appears to be low and it loses out in the competition with copper for the binding sites.[223] Also the binding affinity of zinc to the octarepeat region is much lower than that of Cu^{2+}, with a K_d of about 200 µM.[166] If we compare it with the K_d for the Cu^{2+}–hPrP complex recently proposed by Klewpatinond *et al.* (10^{-8} M),[224] then the affinity of Zn^{2+} towards hPrP is two to three orders of magnitude lower than that of Cu^{2+}.[225] Zinc, even in high concentrations, cannot displace copper from its binding sites in the prion protein, neither from the octarepeats nor from the full-length protein. However, the effect that zinc has on the binding of copper must not be neglected, since it can definitively affect the previously discussed binding modes of Cu^{2+}. It is worth keeping in mind that the physiological concentration of zinc in the brain is significantly higher than that of copper, reaching up to 300 µM. At low Cu^{2+} concentrations, Zn^{2+} can bind to the octarepeat region, and then, when the amount of copper in the brain cell increases, copper will be bound outside the octarepeat region, that is, to His111 and/or His96. Taken together, Zn^{2+}–hPrP interactions could adjust the Cu^{2+} binding modes by directing copper binding to the redox inactive fragment outside the octarepeat region, rather than to the redox accessible octarepeat region.

From the chemical point of view, the binding of copper to this fragment is more effective than the binding of zinc; both metal ions can be sequestered in a $\{4N_{im}\}$ binding mode. For zinc, the mode of binding does not change with increased pH, since it is not able to deprotonate amide nitrogens, contrary to copper complexes where amide nitrogens participate in the binding, leading to a square-planar complex with a $\{N_{im}, 3N^-\}$ donor set.

6.6.1.3 Fe^{2+}

The link between Fe^{2+} and prion disease became obvious after the finding that food-borne transmission of prion infection from cattle (BSE) to humans (CJD) takes place through ferritin, the principal iron storage protein. In epithelial cells of the human intestinal track, PrP^{Sc} was found to be co-transported with ferritin, which suggested that the iron transporter may facilitate the uptake of PrP^{Sc} through the cell barrier across species.[226] Also, a severe Fe^{2+} dyshomeostasis is observed in scrapie-infected cells, leading to abnormally high levels of Fe^{2+}, which, in turn, lead to a significantly increased formation of reactive oxygen species and cell death.[227] The elevated amount of free Fe^{2+} might, however, be a consequence rather than the cause of the disease.

The first and, to the best of our knowledge, the only direct report of Fe^{2+}-induced conversion of PrP^{C} to a PrP^{Sc}-like form was presented by Basu *et al.*,[228] who reported that PrP^{C} binds Fe^{2+} on the surface of human neuroblastoma and transforms it into a PrP^{Sc}-like form, which is similar to the usual pathogenic PrP^{Sc} form, with typical proteinase-K resistance and insolubility. However, the exact mechanism of this conversion remains unknown.

Recently, it has been suggested that the interaction of Fe^{2+} with PrP ought to be neglected, since the greater affinity of iron for other proteins present in the milieu would favour sequestration by them rather than by prion protein.[229,230]

6.6.1.4 Mn^{2+}

Brown *et al.* have claimed that manganese bound to PrP is not displaced when exposed to copper in isothermal titration calorimetric experiments; when the reverse experiment was performed, Mn^{2+} was able to displace Cu^{2+} despite an apparently lower affinity.[231] The replacement of bound Cu^{2+} by Mn^{2+} results in the increase of β-sheet content of PrP^{C}, thereby leading to fibril formation.[223]

Clearly, there must be a link between the binding of Mn^{2+} to PrP and the progression of the disease, since the pathogenic PrP^{Sc} form isolated from the brains of mice with scrapie or patients with CJD has manganese coordinated to it.[232,233]

Also, the levels of Mn^{2+} have been observed to increase approximately 10-fold in brain tissues from sCJD patients.[233] These levels also increase in the blood of sheep, cattle and also in humans directly before the onset of the disease.[234,235]

Biological studies showed that, initially, PrP bound to Mn^{2+} has the same conformation as PrP bound to Cu^{2+},[231] but with time the Mn–PrP complex begins to show typical "scrapie-like" features, such as increased β-sheet content and proteinase K resistance.[223] The changes also occur when Cu^{2+} is still bound to PrP, showing that the loss of copper is unnecessary for the manganese-induced conformational change.[231] Zhu *et al.* recently showed the change in structure that PrP undergoes after binding Mn^{2+}.[236] Also, the neurotoxicity of this metal was studied in mouse neural cells expressing PrP^C and compared with PrP^C knockouts; data from this study proved that the native cellular prion protein protects against Mn^{2+}-induced oxidative stress.[237]

A recent study based on the fragmentary peptide approach showed that the main Mn^{2+} binding site is outside the octarepeat region, at His111 and His96 (or His110 and His95 in the mouse variant of the protein).[238] Another study suggests there is also a second low-affinity Mn^{2+} binding site located in the octarepeat domain, and that the manganese bound to this region becomes irreversibly oxidized.[223]

A detailed NMR and EPR study showed that the human prion protein fragment (PrP106–126, KTNMKHMAGAAAAGAVVGGLG) is able to bind Mn^{2+}, with Gly126, Leu125, Gly124 and His111 residues being involved in the interaction.[238]

From a thermodynamic point of view, the affinity of Mn^{2+} towards PrP is very low when compared to that of Cu^{2+}. However, one should keep in mind that, typically, the affinity of Mn^{2+} towards proteins is rather low, even for those which are manganese binding proteins [the manganese transporter (DMT-1) has a $K_d = 3\,mM$].[239] When we take this into account, the affinity of PrP ($K_d = 63\,\mu M$) is not so small and the debate about whether prions are or are not manganese carriers should definitely continue.

6.6.2   Metals in PrP Aggregation Pathways

With detailed knowledge about the chemical interactions between metal ions and PrP, we can now take a look at such binding on a larger scale and try to answer the question of what happens after metal binding to PrP. Is the consequence always the formation of a neurotoxic form? This usually seems to be the case when more than four copper ions are bound; after the fifth Cu^{2+} binds to PrP^C, the protein undergoes a conformational transition that leads to a highly protease resistant, neurotoxic conformer with an increased β-sheet content (PrPres).[240] These conformers inhibit fibrillization and lead to the precipitation of large aggregates, which are also resistant to proteases.[241,242] Often, the binding of Cu^{2+} to the neurotoxic region results in a folded conformer with a His96–Cu^{2+}–His111 cross-link, which is the starting point of the formation of a β-hairpin that will involve Pro102 and Pro105 at the turn region.[243,244]

The second way in which copper can contribute to the formation of toxic species is by binding to pre-formed prion fibrils, rendering protection against proteolytic degradation.[241] This seems to be confirmed by the finding that

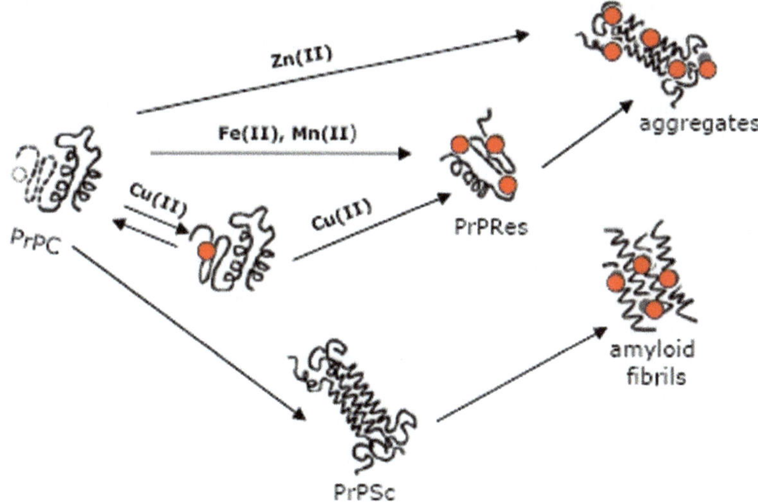

Figure 6.9 Various folding pathways of the human prion protein upon metal binding. The cellular prion protein (PrP^C) can undergo spontaneous misfolding into the PrP^{Sc} form, which templates the misfolding of further proteins and leads to amyloid polymerization. PrP^C can reversibly bind up to four Cu^{2+} ions; when more copper ions are bound, or when Fe^{2+} or Mn^{2+} are available for binding, PrP^C undergoes a conformational change to a β-sheet-rich and protease-resistant conformation (PrPres), which, in turn, forms amorphous aggregates. No intermediate form is known for Zn^{2+} binding.

copper chelation delays the onset of disease in PrP^{Sc} infected mice.[245] On the other hand, copper bound to the unstructured N-terminus can also prevent the establishment of the β-sheet core of the amyloid and can contribute to the non-amyloid aggregation pathway (Figure 6.9).[246]

The binding of Mn^{2+} to PrP, although kinetically slower than that of Cu^{2+},[247] can also induce the formation of PrPres[248] (already at two orders of magnitude lower concentration of Mn^{2+} than that of Cu^{2+}).[249] This transition may, in turn, lead to prion aggregation.[250] Elevated levels of Mn^{2+} are observed in the CNS of CJD patients,[234] and an animal study showed that a Mn^{2+}-specific chelation therapy in individuals infected with prion disease increased their lifespan.[223]

Zn^{2+} and Cu^{2+} ions control the aggregation of the prion protein in a similar way. Provided that the concentration of copper is not abnormally high, both metals have a similar four octarepeat binding mode, when bound 1 : 1 to PrP^C. Like Cu^{2+}, Zn^{2+} can also inhibit the fibrillization of PrP^C by bridging prion molecules and producing neurotoxic aggregates.[222,241]

The binding of Fe^{2+} to PrP induces the formation of the proteinases-resistant form and increases Fe-associated oxidative stress.[251]

In general, as far as the interactions of prion proteins with metal ions are concerned, PrP^C seems to play two basic biological functions: (i) the transport of Cu^{2+} ions and perhaps Zn^{2+}; (ii) *in vivo*, it has been observed that when the prion protein is exposed to metal ions like copper or zinc, the process of internalization of the protein into the cell takes place[252] and it acquires anti-oxidant functions (SOD-like activity). Cu^{2+} ions bound to the prion protein are able to decrease the level of reactive oxygen species, mimicking the activity of copper–zinc superoxide dismutase (CuZn-SOD), an enzyme whose active centre contains a heteronuclear Cu–Zn cluster bridged by the imidazole ring of a histidyl residue.[253] The first suggestion, that prion protein expression may aid cellular resistance to oxidative stress by influencing the activity of Cu/Zn superoxide dismutase, was raised by Brown and Besinger, who showed that increased levels of PrP^C expression were linked to increased levels of CuZn-SOD activity. It was proposed that PrP^C expression may regulate CuZn-SOD activity by influencing the incorporation of copper into the molecule.[169,254] The dismutase function is abolished either when the octarepeat region involved in the copper ion binding is removed or when a copper chelator (for example, diethyldithiocarbamate) is added.

For over a decade this issue was (and still is) a matter of debate. Jones *et al.* showed contradictory results, claiming that the recombinant PrP does not exhibit SOD activity above a detectable threshold.[255] Also, Hutter *et al.* did not detect any SOD activity *in vivo*.[256]

This type of biological activity is also interesting from a chemical point of view. In native SOD the catalytic Cu^{2+} ion is coordinated by three imidazole rings and a water molecule, while the structural Zn^{2+} is bound to two imidazoles and the carboxyl group of an aspartyl residue. The two metal ions are bridged by an additional imidazole residue of His61, whose Nδ and Nε atoms fill the remaining coordination sites of the metal ions. This results in a five-coordinated distorted octahedral geometry for Cu^{2+} and a distorted tetrahedron for Zn^{2+}. Cu^{2+} is crucial for the catalytic SOD activity, while the roles of the imidazolate bridge and Zn^{2+} are less understood and seem to play a different role rather than a catalytic one.

In recent years the SOD-like activity of Cu^{2+}–peptide complexes[257,258] and the SOD-like activity of human, avian, mouse and fish prion fragments with Cu^{2+} have been studied in detail.[27,259,260] Results show that amide coordination has a negative effect on the dismutase activity.[260] At low copper concentrations, when imidazole nitrogens are involved in the metal ion binding, the SOD activity is highest, while its increase results in amide nitrogen coordination and a distinct decrease of this activity. The improvement of SOD activity after addition of Zn^{2+} ions is relatively low compared to that found earlier for ternary Cu^{2+}–Zn^{2+} complexes.[261]

6.7 Neurotransmitters Implicated

The link between prion disease and neurotransmitters is not as pronounced as in the case of Alzheimer's disease or Parkinson's disease, and the question of whether or how the prion protein affects synaptic mechanisms and neuronal excitability remains still open. Since PrP^C is present in both pre- and postsynaptic structures, it might play a role in neuronal communication.[262,263]

PrP^C is more abundant in several regions of the brain (synaptic terminal fields in olfactory bulb, limbic-associated structures and in the striato-nigral complex) than in others (fibre pathways, with the exception of the olfactory nerve).[264] One of the earliest studies on PrP showed that, in the case of scrapie, even though histological lesions are widespread in the CNS, no consistent neurochemical changes were found in mid- or anterior brain. A reduced activity of glutamic acid decarboxylase was found whereas 5-hydroxyindoleacetic acid (5-HIAA) increased, while the levels of serotonin and dopamine, as well as the binding of ligands to serotonin and to musearinic cholinergic receptors, decreased. In the spinal cord, the activity of choline acetyltransferase was reduced and the concentration of 5-HIAA was more pronounced.[265]

Later reports confirmed these studies and provided insight into the role of the native prion protein. Recently, it has been shown that PrP^C plays a role in dopamine metabolism by regulating monoamine oxidase (MAO) activity (MAO is involved in oxidative degradation of dopamine)[266] and that PrP^C influences synaptic transmission by potentiating acetylcholine release at the neuromuscular junction.[267]

Another way in which PrP^C may affect signal transduction in synapses is connected with the fact that PrP^C is able to bind copper: because nerve endings release copper into the synaptic cleft upon depolarization, it is probable that the presynaptic PrP^C may buffer Cu^{2+} levels in the synaptic cleft and ensure its transport back into the presynaptic cytosol. This mechanism could, at the same time, protect synapses from reactive oxygen species generated by Fenton-type redox reactions.[268] Such buffering of Cu^{2+} by PrP^C could play a role in calcium homeostasis, because in the synaptic cleft, copper is able to reduce Ca^{2+} influx through voltage-gated calcium channels.[269]

There is convincing evidence for a role of PrP^C in behavioural processes, particularly in memory. "Long-term potentiation" (LTP; a long-lasting enhancement in signal transmission between two neurons that results from stimulating them synchronously) has long been associated with learning and memory. In several studies, impaired LTP was found in tissue slices from PrP-null mice, when compared with wild type.[270] Apart from this, no deficits were found in cell excitability, synaptic inhibition, reversal potential for inhibitory postsynaptic potentials or LTP.

Another recent animal study showed that when PrP^C is absent, the excitatory mechanisms are altered and epileptic seizures may occur. What is surprising is that greater amounts of PrP^C do not provide greater protection against seizures, but have the opposite effect, increasing the possibility of suffering severe epileptic seizures. The level of excitability of the CNS is increased even more than in the absence of PrP^C, due to the fact that both the excitatory and inhibitory mechanisms are altered.[271]

This is not the only case in which both a deficiency and an excess of PrP have a considerable effect on neurotransmitter homeostasis. Both the absence and the overexpression of the prion protein affect the levels of expression of glutamate receptors (AMPA-kainate, an excitatory neurotransmitter) and of GABA (an inhibitory neurotransmitter).[271]

These data lead to the conclusion that adequate concentrations of PrP, by modulating neuronal excitability and synaptic activity, contribute to maintaining a balance between the mechanisms that excite neurons and the mechanisms that inhibit them. Although those processes are not yet entirely understood, it can be concluded that prion proteins play a role in maintaining neurotransmitter homeostasis in the CNS.

6.8 Therapeutics

Many compounds have been tested as potential anti-prion agents. Many of them showed distinct inhibition of prion propagation in the development of the disease.[272,273] However, just before or during onset they were shown to be ineffective.[274,275]

Using the scrapie infected neuroblastoma cells (ScN2a), Prusiner *et al.* have tested a series of derivatives of acridine and phenothiazine as potential therapeutics for prion disease.[276] Both families of compounds penetrate the BBB, which is a critical property needed for potential drugs. Phenothiazines, a family of compounds comprising a tricyclic scaffold, had already been used for decades to treat psychosis.[277] Using a PrP^{Sc} model (ScN2a cells),[276] a series of analogues were shown to inhibit PrP^{Sc} formation and enhance clearance of scrapie prions from the cells. Many of the tested compounds had already been used previously as drugs, *e.g.* against malaria.[278] Some derivatives of phenothiazine, like chlorpromazine and acepromazine, were found to be quite efficient in reducing PrP^{Sc} in ScN2a cells. Many phenothiazines, however, were shown to be very cytotoxic. Quinacrine, an acridine derivative and antimalarial drug, on the other hand, was found to be much better than chlorpromazine and a very potent inhibitor of PrP^{Sc} formation. It was suggested that quinacrine is a good candidate for the lead anti-prion compound. The clinical studies showed that quinacrine at a dose of 300 mg per day is reasonably tolerated by patients but did not significantly affect prion disease development.[279] It is possible that the low dose of quinacrine used in this study was the reason for the lack of therapeutic effects.

Polyanions have been shown to inhibit progression of spongiform encephalopathies.[280,281] Sulfated polysaccharides are rather effective in inhibiting the formation of PrP^{Sc} *in vitro*[282] and in some peripherally infected animal models.[283–285] Pentosan polysulfate (PPS) is one of the most effective polyanions, with a heparin-like structure. However, it does not cross the BBB after oral or parenteral administration.[286] PPS injected directly into the cerebral ventricle of prion-infected animals is quite effective in inhibiting formation of PrP^{Sc}.[287] This result inspired the introduction of intraventricular infusions of PPS into human brain. Prion diseases are rare and available clinical data are insufficient to draw general conclusions. The UK study[288] has shown that there are rather frequent complications of intraventricular catheterization, although PPS was well tolerated within a wide range of doses and the mean survival of all patients was longer than expected for this type of disease. In the case of Japanese studies[289] it was shown that continuous intraventricular infusion caused no apparent improvement; however, the possibility of long-term treatment was suggested. Thus, even if the toxicity of PPS is moderate, we still need real proof that it is effective against neurodegenerative disorders.[290,291] Flupritine maleate (a triaminopyridine compound) is well tolerated, displaying cytoprotective activity *in vitro* and *in vivo*. However, as an anti-neurodegenerative drug it needs further studies both in animal models and in a clinical context.[292]

Organic molecules which have been examined as potential drugs for neuro-degenerative disorders are often derived from known pharmacologically active families of chemical compounds used previously in different diseases or at least tested as potential drugs for other pathologies (Figure 6.10) (*vide supra*). The possible involvement of metal ions in neurodegenerative pathologies has induced intensive research on the impact of metal ions on protein aggregation processes or oxidative stress accompanying neurode-generation.[26,27,229,230,293–295] As metal ions could have a critical impact on protein aggregation and are directly involved in the production of oxygen radicals (oxidative stress), the natural solution for these problems are chelating agents regulating metal homeostasis usually disturbed during the neuro-degeneration process, even if the role of metal ions is still unclear.

Chelating agents with potential application for neurodegenerative disorders should satisfy several basic criteria, such as important lipophilicity to cross not only cellular membranes but also the BBB, low toxicity, rather low molecular weight, and not too high affinity for metal ions to protect against complete removal of metal ions from the brain or other organs. As neurodegenerative disorders are very complex with complicated pathology, simple drugs inter-acting with a single target are usually insufficient. Modern strategies try to include more than one function into a single molecule to direct it to more than one target involved in the neurodegenerative pathology.[296–303]

There are different ways to combine several functions in one molecule. Chelating agents could be one of the elements in such multifunctional drug molecules. Metal ions are one of the targets which should be considered in designing drugs.[304] Zn^{2+} and Cu^{2+} ions are thought to be involved in the aggregation process of, for example, β-amyloid peptide (Alzheimer's disease) and many promising results have been obtained from *in vitro* studies showing that chelators like EDTA (ethylenediaminetetraacetic acid) or DTPA (diethylenetriaminepentaacetic acid) were able to reverse peptide aggre-gation.[305] However, these chelators cross the BBB poorly and their use for neurologic problems is less effective. Clioquinol (5-chloro-7-iodoquinolin-8-ol, CQ), which is rather small and lipophilic and not a very good chelator of Zn^{2+} and Cu^{2+}, was found to cross the BBB well. It was approved by the FDA, although clinical trials were stopped after some difficulties with its synthesis and toxicity.[306,307] Although clinical trials of CQ were discontinued, the medical results obtained have inspired large-scale studies on various analogues and derivatives.

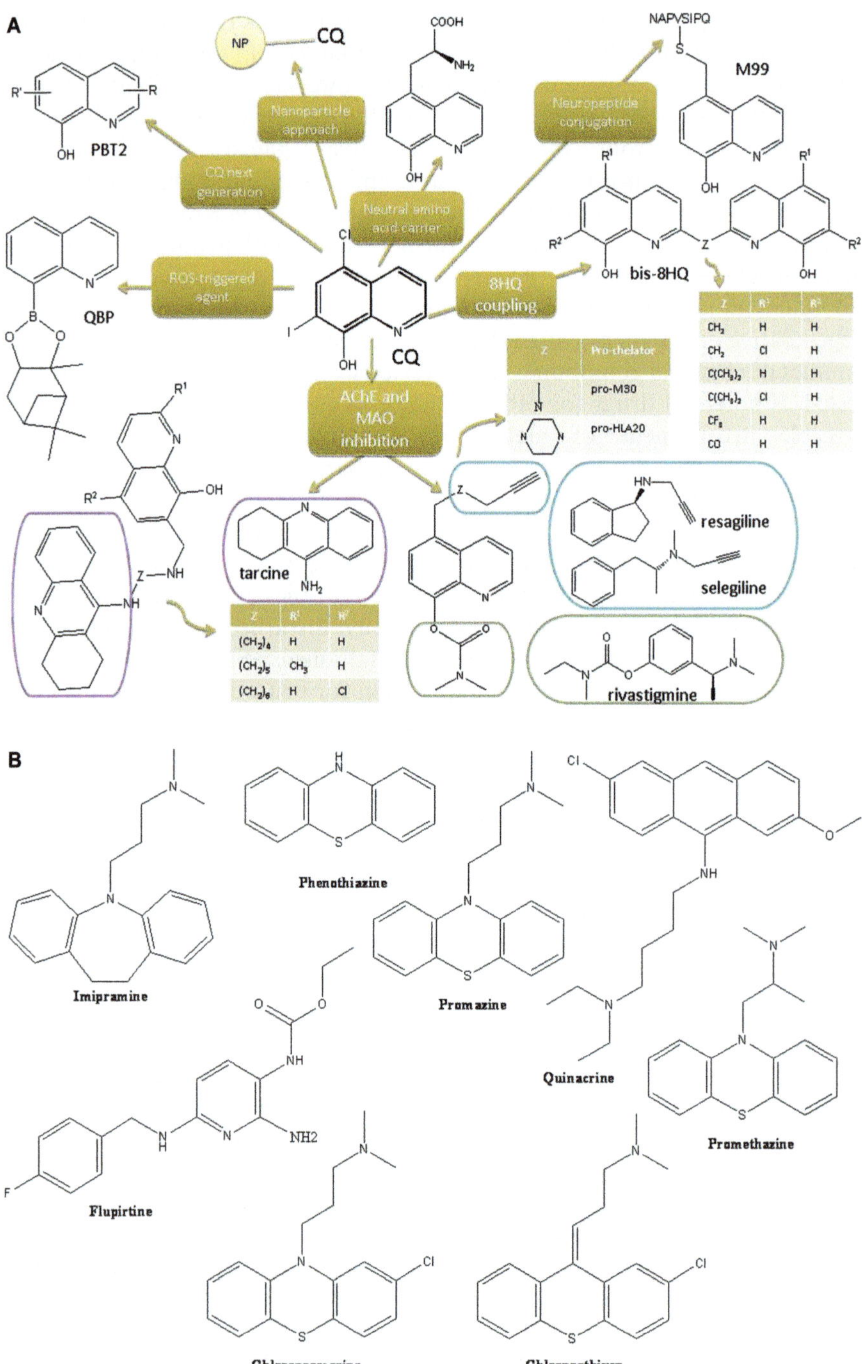

Figure 6.10 (A) Multifunctional ligands based on clioquinol (Adapted from Rodriguez-Rodriguez *et al.*[293]). (B) Different drugs used in the treatment of prion diseases.

References

1. C. Soto, *Trends Biochem. Sci.*, 2011, **36**, 151.
2. H. Budka, *Br. Med. Bull.*, 2003, **66**, 121.
3. G. G. Kovacs and H. Budka, *Int. J. Mol. Sci.*, 2009, **10**, 976.
4. C. Soto and N. Satani, *Trends Mol. Med.*, 2011, **17**, 14.
5. S. B. Prusiner, *Science*, 1982, **216**, 136.
6. A. Aguzzi and A. M. Calella, *Physiol. Rev.*, 2009, **89**, 1105.
7. S. B. Prusiner, *Proc. Natl. Acad. Sci. U. S. A.*, 1998, **95**, 13363.
8. L. Westergard, H. M. Christensen and D. A. Harris, *Biochim. Biophys. Acta*, 2007, **1772**, 629.
9. L. B. Chiarini, A. R. O. Freitas, S. M. Zanata, R. R. Brentani, V. R. Martins and R. Linden, *EMBO J.*, 2002, **21**, 3317.
10. B. Hugel, M. Martinez, C. Kunzelmann, T. Blattler, A. Aguzzi and J. M. Freyssinet, *Cell. Mol. Life Sci.*, 2004, **61**, 2998.
11. A. Sakudo and K. Ikuta, *Protein Pept. Lett*, 2009, **16**, 217.
12. D. R. Brown, *Biochem. Soc. Trans.*, 2002, **30**, 742.
13. D. R. Brown and H. Kozlowski, *Dalton Trans.*, 2004, 1907.
14. X. Roucou and A. C. LeBlanc, *J. Mol. Med.*, 2005, **83**, 3.
15. C. Hetz, K. Maundrell and C. Soto, *Trends Mol. Med.*, 2003, **9**, 237.
16. C. Soto, *Nat. Rev. Neurosci.*, 2003, **4**, 49.
17. C. I. Lasmezas, J. P. Deslys, O. Robain, A. Jaegly, V. Beringue, J. M. Peyrin, J. G. Fournier, J. J. Hauw, J. Rossier and D. Dormont, *Science*, 1997, **275**, 402.
18. P. Piccardo, J. C. Manson, D. King, B. Ghetti and R. M. Barron, *Proc. Natl. Acad. Sci. U. S. A.*, 2007, **104**, 4712.
19. E. Biasini, J. A. Turnbaugh, U. Unterberger and D. A. Harris, *Trends Neurosci.*, 2012, **35**, 92.
20. S. Tzaban, G. Friedlander, O. Schonberger, L. Horonchik, Y. Yedidia, G. Shaked, R. Gabizon and A. Taraboulos, *Biochemistry*, 2002, **41**, 12868.
21. J. R. Silveira, G. J. Raymond, A. G. Hughson, R. E. Race, V. L. Sim, S. F. Hayes and B. Caughey, *Nature*, 2005, **437**, 257.
22. P. Gambetti, Z. Dong, J. Yuan, X. Xiao, M. Zheng, A. Alshekhlee, R. Castellani, M. Cohen, M. A. Barria, D. Gonzalez-Romero, E. D. Belay, L. B. Schonberger, K. Marder, C. Harris, J. R. Burke, T. Montine, T. Wisniewski, D. W. Dickson, C. Soto, C. M. Hulette, J. A. Mastrianni, Q. Z. Kong and W. Q. Zou, *Ann. Neurol.*, 2008, **63**, 697.
23. D. W. Colby, K. Giles, G. Legname, H. Wille, I. V. Baskakov, S. J. DeArmond and S. B. Prusiner, *Proc. Natl. Acad. Sci. U. S. A.*, 2009, **106**, 20417.
24. G. Legname, I. V. Baskakov, H. O. B. Nguyen, D. Riesner, F. E. Cohen, S. J. DeArmond and S. B. Prusiner, *Science*, 2004, **305**, 673.
25. G. Legname, H. O. B. Nguyen, I. V. Baskakov, F. E. Cohen, S. J. DeArmond and S. B. Prusiner, *Proc. Natl. Acad. Sci. U. S. A.*, 2005, **102**, 2168.

26. E. Gaggelli, H. Kozlowski, D. Valensin and G. Valensin, *Chem. Rev.*, 2006, **106**, 1995.

27. H. Kozlowski, D. R. Brown and G. Valensin, *Metallochemistry of Neurodegeneration: Biological, Chemical, and Genetic Aspects*, Royal Society of Chemistry, Cambridge, 2006.

28. H. Budka, A. Aguzzi, P. Brown, J. M. Brucher, O. Bugiani, F. Gullotta, M. Haltia, J. J. Hauw, J. W. Ironside, K. Jellinger, H. A. Kretzschmar, P. L. Lantos, C. Masullo, W. Schlote, J. Tateishi and R. O. Weller, *Brain Pathol.*, 1995, **5**, 459.

29. B. Ghetti, S. R. Dlouhy, G. Giaccone, O. Bugiani, B. Frangione, M. R. Farlow and F. Tagliavini, *Brain Pathol.*, 1995, **5**, 61.

30. M. D. Spencer, R. S. G. Knight and R. G. Will, *Br. Med. J.*, 2002, **324**, 1479.

31. J. Collinge, *Annu. Rev. Neurosci.*, 2001, **24**, 519.

32. J. Collinge, *J. Neurol. Neurosurg. Psychiat.*, 2005, **76**, 906.

33. A. F. Hill, S. Joiner, J. A. Beck, T. A. Campbell, A. Dickinson, M. Poulter, J. D. F. Wadsworth and J. Collinge, *Brain*, 2006, **129**, 676.

34. G. G. Kovacs, G. Trabattoni, J. A. Hainfellner, J. W. Ironside, R. S. G. Knight and H. Budka, *J. Neurol.*, 2002, **249**, 1567.

35. S. Mead, *Eur. J. Hum. Genet.*, 2006, **14**, 273.

36. J. D. Wadsworth, A. F. Hill, J. A. Beck and J. Collinge, *Br. Med. Bull.*, 2003, **66**, 241.

37. S. E. Lloyd and J. Collinge, *Curr. Genomics*, 2005, **6**, 1.

38. S. Capellari, R. Strammiello, D. Saverioni, H. Kretzschmar and P. Parchi, *Acta Neuropathol.*, 2011, **121**, 21.

39. L. Cervenakova, L. G. Goldfarb, R. Garruto, H. S. Lee, D. C. Gajdusek and P. Brown, *Proc. Natl. Acad. Sci. U. S. A.*, 1998, **95**, 13239.

40. H. F. Baker, M. Poulter, T. J. Crow, C. D. Frith, R. Lofthouse, R. M. Ridley and J. Collinge, *Lancet*, 1991, **337**, 1286.

41. J. Collinge, J. Brown, J. Hardy, M. Mullan, M. N. Rossor, H. Baker, T. J. Crow, R. Lofthouse, M. Poulter, R. Ridley, F. Owen, C. Bennett, G. Dunn, A. E. Harding, N. Quinn, B. Doshi, G. W. Roberts, M. Honavar, I. Janota and P. L. Lantos, *Brain*, 1992, **115**, 687.

42. J. Collinge, M. S. Palmer and A. J. Dryden, *Lancet*, 1991, **337**, 1441.

43. M. S. Palmer, A. J. Dryden, J. T. Hughes and J. Collinge, *Nature*, 1991, **352**, 340.

44. B. Drisaldi, R. S. Stewart, C. Adles, L. R. Stewart, E. Quaglio, E. Biasini, L. Fioriti, R. Chiesa and D. A. Harris, *J. Biol. Chem.*, 2003, **278**, 21732.

45. L. Ivanova, S. Barmada, T. Kummer and D. A. Harris, *J. Biol. Chem.*, 2001, **276**, 42409.

46. J. Y. Ma and S. Lindquist, *Proc. Natl. Acad. Sci. U. S. A.*, 2001, **98**, 14955.

47. A. Ashok and R. S. Hegde, *PLoS Pathog.*, 2009, **5**, e1000479.

48. V. Campana, D. Sarnataro, C. Fasano, P. Casanova, S. Paladino and C. Zurzolo, *J. Cell Sci.*, 2006, **119**, 433.

49. T. Massignan, E. Biasini, E. Lauranzano, P. Veglianese, M. Pignataro, L. Fioriti, D. A. Harris, M. Salmona, R. Chiesa and V. Bonetto, *Mol. Cell. Proteomics*, 2010, **9**, 611.

50. E. Schiff, V. Campana, S. Tivodar, S. Lebreton, K. Gousset and C. Zurzolo, *Traffic*, 2008, **9**, 1101.

51. S. G. Chen, P. Parchi, P. Brown, S. Capellari, W. Q. Zou, E. J. Cochran, C. L. Vnencak-Jones, J. Julien, C. Vital, J. Mikol, E. Lugaresi, L. Autilio-Gambetti and P. Gambetti, *Nat. Med.*, 1997, **3**, 1009.

52. R. B. Petersen, P. Parchi, S. L. Richardson, C. B. Urig and P. Gambetti, *J. Biol. Chem.*, 1996, **271**, 12661.

53. W. S. Jackson, A. W. Borkowski, H. Faas, A. D. Steele, O. D. King, N. Watson, A. Jasanoff and S. Lindquist, *Neuron*, 2009, **63**, 438.

54. P. Parchi, S. Capellari, S. G. Chen, R. B. Petersen, P. Gambetti, N. Kopp, P. Brown, T. Kitamoto, J. Tateishi, A. Giese and H. Kretzschmar, *Nature*, 1997, **386**, 232.

55. P. Parchi, R. Castellani, S. Capellari, B. Ghetti, K. Young, S. G. Chen, M. Farlow, D. W. Dickson, A. A. F. Sima, J. Q. Trojanowski, R. B. Petersen and P. Gambetti, *Ann. Neurol.*, 1996, **39**, 767.

56. P. Parchi, W. Q. Zou, W. Wang, P. Brown, S. Capellari, B. Ghetti, N. Kopp, W. J. Schulz-Schaeffer, H. A. Kretzschmar, M. W. Head, J. W. Ironside, P. Gambetti and S. G. Chen, *Proc. Natl. Acad. Sci. U. S. A.*, 2000, **97**, 10168.

57. G. Tamguney, K. Giles, D. V. Glidden, P. Lessard, H. Wille, P. Tremblay, D. F. Groth, F. Yehiely, C. Korth, R. C. Moore, J. Tatzelt, E. Rubinstein, C. Boucheix, X. P. Yang, P. Stanley, M. P. Lisanti, R. A. Dwek, P. M. Rudd, J. Moskovitz, C. J. Epstein, T. D. Cruz, W. A. Kuziel, N. Maeda, J. Sap, K. H. Ashe, G. A. Carlson, I. Tesseur, T. Wyss-Coray, L. Mucke, K. H. Weisgraber, R. W. Mahley, F. E. Cohen and S. B. Prusiner, *J. Gen. Virol.*, 2008, **89**, 1777.

58. G. Schmitt-Ulms, K. Hansen, J. L. Liu, C. Cowdrey, J. Yang, S. J. DeArmond, F. E. Cohen, S. B. Prusiner and M. A. Baldwin, *Nat. Biotechnol.*, 2004, **22**, 724.

59. L. M. Felton, C. Cunningham, E. L. Rankine, S. Waters, D. Boche and V. H. Perry, *Neurobiol. Dis.*, 2005, **20**, 283.

60. C. Soto and J. Castilla, *Nat. Med.*, 2004, **10**(suppl), S63.

61. K. W. Leffers, H. Wille, J. Stohr, E. Junger, S. B. Prusiner and D. Riesner, *Biol. Chem.*, 2005, **386**, 569.

62. S. B. Prusiner, D. F. Groth, D. C. Bolton, S. B. Kent and L. E. Hood, *Cell*, 1984, **38**, 127.

63. C. Govaerts, H. Wille, S. B. Prusiner and F. E. Cohen, *Proc. Natl. Acad. Sci. U. S. A.*, 2004, **101**, 8342.

64. X. Lu, P. L. Wintrode and W. K. Surewicz, *Proc. Natl. Acad. Sci. U. S. A.*, 2007, **104**, 1510.

65. N. J. Cobb, F. D. Sonnichsen, H. McHaourab and W. K. Surewicz, *Proc. Natl. Acad. Sci. U. S. A.*, 2007, **104**, 18946.

66. N. Stahl, D. R. Borchelt, K. Hsiao and S. B. Prusiner, *Cell*, 1987, **51**, 229.

67. P. M. Rudd, M. R. Wormald, D. R. Wing, S. B. Prusiner and R. A. Dwek, *Biochemistry*, 2001, **40**, 3759.
68. B. Vincent, E. Paitel, P. Saftig, Y. Frobert, D. Hartmann, B. De Strooper, J. Grassi, E. Lopez-Perez and F. Checler, *J. Biol. Chem.*, 2001, **276**, 37743.
69. S. G. Chen, D. B. Teplow, P. Parchi, J. K. Teller, P. Gambetti and L. Autilio-Gambetti, *J. Biol. Chem.*, 1995, **270**, 19173.
70. F. Wopfner, G. Weidenhofer, R. Schneider, A. von Brunn, S. Gilch, T. F. Schwarz, T. Werner and H. M. Schatzl, *J. Mol. Biol.*, 1999, **289**, 1163.
71. E. Rivera-Milla, C. A. Stuermer and E. Malaga-Trillo, *Trends Genet.*, 2003, **19**, 72.
72. B. Strumbo, S. Ronchi, L. C. Bolis and T. Simonic, *FEBS Lett.*, 2001, **508**, 170.
73. K. Wuthrich and R. Riek, *Adv. Protein Chem.*, 2001, **57**, 55.
74. R. Zahn, A. Liu, T. Luhrs, R. Riek, C. von Schroetter, F. Lopez Garcia, M. Billeter, L. Calzolai, G. Wider and K. Wuthrich, *Proc. Natl. Acad. Sci. U. S. A.*, 2000, **97**, 145.
75. R. S. Hegde, J. A. Mastrianni, M. R. Scott, K. A. DeFea, P. Tremblay, M. Torchia, S. J. DeArmond, S. B. Prusiner and V. R. Lingappa, *Science*, 1998, **279**, 827.
76. R. Riek, S. Hornemann, G. Wider, M. Billeter, R. Glockshuber and K. Wuthrich, *Nature*, 1996, **382**, 180.
77. F. F. Damberger, B. Christen, D. R. Perez, S. Hornemann and K. Wuthrich, *Proc. Natl. Acad. Sci. U. S. A.*, 2011, **108**, 17308.
78. A. Pastore and A. Zagari, *Prion*, 2007, **1**, 185.
79. S. Hornemann, C. Schorn and K. Wuthrich, *EMBO Rep.*, 2004, **5**, 1159.
80. H. Eberl, P. Tittmann and R. Glockshuber, *J. Biol. Chem.*, 2004, **279**, 25058.
81. B. Oesch, D. Westaway, M. Walchli, M. P. McKinley, S. B. Kent, R. Aebersold, R. A. Barry, P. Tempst, D. B. Teplow, L. E. Hood, *et al.*, *Cell*, 1985, **40**, 735.
82. E. Biasini, J. A. Turnbaugh, U. Unterberger and D. A. Harris, *Trends Neurosci.*, 2012, **35**, 92.
83. H. Bueler, M. Fischer, Y. Lang, H. Bluethmann, H. P. Lipp, S. J. DeArmond, S. B. Prusiner, M. Aguet and C. Weissmann, *Nature*, 1992, **356**, 577.
84. J. C. Manson, A. R. Clarke, M. L. Hooper, L. Aitchison, I. McConnell and J. Hope, *Mol. Neurobiol.*, 1994, **8**, 121.
85. C. E. Le Pichon, M. T. Valley, M. Polymenidou, A. T. Chesler, B. T. Sagdullaev, A. Aguzzi and S. Firestein, *Nat. Neurosci.*, 2009, **12**, 60.
86. J. Bremer, F. Baumann, C. Tiberi, C. Wessig, H. Fischer, P. Schwarz, A. D. Steele, K. V. Toyka, K. A. Nave, J. Weis and A. Aguzzi, *Nat. Neurosci.*, 2010, **13**, 310.
87. A. D. Steele, S. Lindquist and A. Aguzzi, *Prion*, 2007, **1**, 83.
88. R. C. Moore, I. Y. Lee, G. L. Silverman, P. M. Harrison, R. Strome, C. Heinrich, A. Karunaratne, S. H. Pasternak, M. A. Chishti, Y. Liang,

P. Mastrangelo, K. Wang, A. F. Smit, S. Katamine, G. A. Carlson, F. E. Cohen, S. B. Prusiner, D. W. Melton, P. Tremblay, L. E. Hood and D. Westaway, *J. Mol. Biol.*, 1999, **292**, 797.

89. E. Rivera-Milla, B. Oidtmann, C. H. Panagiotidis, M. Baier, T. Sklaviadis, R. Hoffmann, Y. Zhou, G. P. Solis, C. A. Stuermer and E. Malaga-Trillo, *FASEB J.*, 2006, **20**, 317.

90. J. Kanaani, S. B. Prusiner, J. Diacovo, S. Baekkeskov and G. Legname, *J. Neurochem.*, 2005, **95**, 1373.

91. M. Fuhrmann, T. Bittner, G. Mitteregger, N. Haider, S. Moosmang, H. Kretzschmar and J. Herms, *J. Neurochem.*, 2006, **98**, 1876.

92. C. Lazzari, C. Peggion, R. Stella, M. L. Massimino, D. Lim, A. Bertoli and M. C. Sorgato, *J. Neurochem.*, 2011, **116**, 881.

93. F. Prestori, P. Rossi, B. Bearzatto, J. Laine, D. Necchi, S. Diwakar, S. N. Schiffmann, H. Axelrad and E. D'Angelo, *J. Neurosci.*, 2008, **28**, 7091.

94. I. H. Solomon, N. Khatri, E. Biasini, T. Massignan, J. E. Huettner and D. A. Harris, *J. Biol. Chem.*, 2011, **286**, 14724.

95. R. Linden, V. R. Martins, M. A. Prado, M. Cammarota, I. Izquierdo and R. R. Brentani, *Physiol. Rev.*, 2008, **88**, 673.

96. E. Graner, A. F. Mercadante, S. M. Zanata, O. V. Forlenza, A. L. Cabral, S. S. Veiga, M. A. Juliano, R. Roesler, R. Walz, A. Minetti, I. Izquierdo, V. R. Martins and R. R. Brentani, *Brain Res. Mol. Brain Res.*, 2000, **76**, 85.

97. R. Rieger, F. Edenhofer, C. I. Lasmezas and S. Weiss, *Nat. Med.*, 1997, **3**, 1383.

98. S. Gauczynski, J. M. Peyrin, S. Haik, C. Leucht, C. Hundt, R. Rieger, S. Krasemann, J. P. Deslys, D. Dormont, C. I. Lasmezas and S. Weiss, *EMBO J.*, 2001, **20**, 5863.

99. G. Schmitt-Ulms, G. Legname, M. A. Baldwin, H. L. Ball, N. Bradon, P. J. Bosque, K. L. Crossin, G. M. Edelman, S. J. DeArmond, F. E. Cohen and S. B. Prusiner, *J. Mol. Biol.*, 2001, **314**, 1209.

100. V. Devanathan, I. Jakovcevski, A. Santuccione, S. Li, H. J. Lee, E. Peles, I. Leshchyns'ka, V. Sytnyk and M. Schachner, *J. Neurosci.*, 2010, **30**, 9292.

101. E. Malaga-Trillo, G. P. Solis, Y. Schrock, C. Geiss, L. Luncz, V. Thomanetz and C. A. Stuermer, *PLoS Biol.*, 2009, **7**, e55.

102. W. C. Shyu, S. Z. Lin, M. F. Chiang, D. C. Ding, K. W. Li, S. F. Chen, H. I. Yang and H. Li, *J. Neurosci.*, 2005, **25**, 8967.

103. B. H. Kim, H. G. Lee, J. K. Choi, J. I. Kim, E. K. Choi, R. I. Carp and Y. S. Kim, *Brain Res. Mol. Brain Res.*, 2004, **124**, 40.

104. C. L. Haigh, S. C. Drew, M. P. Boland, C. L. Masters, K. J. Barnham, V. A. Lawson and S. J. Collins, *J. Cell Sci.*, 2009, **122**, 1518.

105. S. M. Zanata, M. H. Lopes, A. F. Mercadante, G. N. Hajj, L. B. Chiarini, R. Nomizo, A. R. Freitas, A. L. Cabral, K. S. Lee, M. A. Juliano, E. de Oliveira, S. G. Jachieri, A. Burlingame, L. Huang, R. Linden, R. R. Brentani and V. R. Martins, *EMBO J.*, 2002, **21**, 3307.

106. T. G. Santos, I. R. Silva, B. Costa-Silva, A. P. Lepique, V. R. Martins and M. H. Lopes, *Stem Cells*, 2011, **29**, 1126.
107. A. S. Rambold, V. Muller, U. Ron, N. Ben-Tal, K. F. Winklhofer and J. Tatzelt, *EMBO J.*, 2008, **27**, 1974.
108. J. C. Watts and D. Westaway, *Biochim. Biophys. Acta*, 2007, **1772**, 654.
109. A. Behrens, N. Genoud, H. Naumann, T. Rulicke, F. Janett, F. L. Heppner, B. Ledermann and A. Aguzzi, *EMBO J.*, 2002, **21**, 3652.
110. S. Sakaguchi, S. Katamine, N. Nishida, R. Moriuchi, K. Shigematsu, T. Sugimoto, A. Nakatani, Y. Kataoka, T. Houtani, S. Shirabe, H. Okada, S. Hasegawa, T. Miyamoto and T. Noda, *Nature*, 1996, **380**, 528.
111. N. Daude and D. Westaway, *Front. Biosci.*, 2011, **16**, 1505.
112. M. Premzl, L. Sangiorgio, B. Strumbo, J. A. Marshall Graves, T. Simonic and J. E. Gready, *Gene*, 2003, **314**, 89.
113. D. C. Gajdusek, *Science*, 1977, **197**, 943.
114. S. B. Prusiner, *Science*, 1982, **216**, 136.
115. G. P. Saborio, B. Permanne and C. Soto, *Nature*, 2001, **411**, 810.
116. P. Meier, N. Genoud, M. Prinz, M. Maissen, T. Rulicke, A. Zurbriggen, A. J. Raeber and A. Aguzzi, *Cell*, 2003, **113**, 49.
117. C. Weissmann, *Nature*, 1991, **349**, 569.
118. B. Chesebro, M. Trifilo, R. Race, K. Meade-White, C. Teng, R. LaCasse, L. Raymond, C. Favara, G. Baron, S. Priola, B. Caughey, E. Masliah and M. Oldstone, *Science*, 2005, **308**, 1435.
119. K. Abid and C. Soto, *Cell. Mol. Life Sci.*, 2006, **63**, 2342.
120. N. R. Deleault, R. W. Lucassen and S. Supattapone, *Nature*, 2003, **425**, 717.
121. N. R. Deleault, J. C. Geoghegan, K. Nishina, R. Kascsak, R. A. Williamson and S. Supattapone, *J. Biol. Chem.*, 2005, **280**, 26873.
122. F. Wang, X. Wang, C. G. Yuan and J. Ma, *Science*, 2010, **327**, 1132.
123. G. S. Baron, A. C. Magalhaes, M. A. Prado and B. Caughey, *J. Virol.*, 2006, **80**, 2106.
124. S. Liemann and R. Glockshuber, *Biochemistry*, 1999, **38**, 3258.
125. W. Swietnicki, R. B. Petersen, P. Gambetti and W. K. Surewicz, *J. Biol. Chem.*, 1998, **273**, 31048.
126. G. C. Telling, M. Scott, J. Mastrianni, R. Gabizon, M. Torchia, F. E. Cohen, S. J. DeArmond and S. B. Prusiner, *Cell*, 1995, **83**, 79.
127. K. Kaneko, L. Zulianello, M. Scott, C. M. Cooper, A. C. Wallace, T. L. James, F. E. Cohen and S. B. Prusiner, *Proc. Natl. Acad. Sci. U. S. A.*, 1997, **94**, 10069.
128. A. Ashok and R. S. Hegde, *PLoS Pathog.*, 2009, **5**, e1000479.
129. G. Rossetti, G. Giachin, G. Legname and P. Carloni, *Proteins*, 2010, **78**, 3270.
130. J. T. Jarrett and P. T. Lansbury, Jr., *Cell*, 1993, **73**, 1055.
131. P. T. Lansbury, *Science*, 1994, **265**, 1510.
132. S. B. Prusiner, M. P. McKinley, K. A. Bowman, D. C. Bolton, P. E. Bendheim, D. F. Groth and G. G. Glenner, *Cell*, 1983, **35**, 349.

133. S. B. Prusiner, D. C. Bolton, D. F. Groth, K. A. Bowman, S. P. Cochran and M. P. McKinley, *Biochemistry*, 1982, **21**, 6942.
134. P. L. McGeer and E. G. McGeer, *Neurobiol. Aging*, 2001, **22**, 799.
135. A. E. Williams, L. J. Lawson, V. H. Perry and H. Fraser, *Neuropath. Appl. Neurobiol.*, 1994, **20**, 47.
136. V. H. Perry, C. Cunningham and D. Boche, *Curr. Opin. Neurol.*, 2002, **15**, 349.
137. P. Eikelenboom, C. Bate, W. A. Van Gool, J. J. M. Hoozemans, J. M. Rozemuller, R. Veerhuis and A. Williams, *Glia*, 2002, **40**, 232.
138. F. Aloisi, F. Ria and L. Adorini, *Immunol. Today*, 2000, **21**, 141.
139. D. R. Brown, *Microsc. Res. Tech.*, 2001, **54**, 71.
140. S. J. McHattie, D. R. Brown and M. M. Bird, *J. Neurocytol.*, 1999, **28**, 149.
141. V. Beringue, M. Demoy, C. I. Lasmezas, B. Gouritin, C. Weingarten, J. P. Deslys, J. P. Andreux, P. Couvreur and D. Dormont, *J. Pathol.*, 2000, **190**, 495.
142. A. M. Thackray, A. N. McKenzie, M. A. Klein, A. Lauder and R. Bujdoso, *J. Virol.*, 2004, **78**, 13697.
143. T. Kielian, *Front. Biosci.*, 2004, **9**, 732.
144. A. Szczucinski and J. Losy, *Acta Neurol. Scand.*, 2007, **115**, 137.
145. I. Tsunoda, T. E. Lane, J. Blackett and R. S. Fujinami, *Mult. Scler.*, 2004, **10**, 26.
146. Q. Wang, X. N. Tang and M. A. Yenari, *J. Neuroimmunol.*, 2007, **184**, 53.
147. P. Eikelenboom, R. Veerhuis, W. Scheper, A. J. M. Rozemuller, W. A. van Gool and J. J. M. Hoozemans, *J. Neural Transm.*, 2006, **113**, 1685.
148. T. Wyss-Coray, *Nat. Med.*, 2006, **12**, 1005.
149. R. M. Ransohoff, L. Liu and A. E. Cardona, *Int. Rev. Neurobiol.*, 2007, **82**, 187.
150. C. Savarin-Vuaillat and R. M. Ransohof, *Neurotherapeutics*, 2007, **4**, 590.
151. C. A. Baker, D. Martin and L. Manuelidis, *J. Virol.*, 2002, **76**, 10905.
152. M. Burwinkel, C. Riemer, A. Schwarz, J. Schultz, S. Neidhold, T. Bamme and M. Baier, *Int. J. Dev. Neurosci.*, 2004, **22**, 497.
153. C. Riemer, I. Queck, D. Simon, R. Kurth and M. Baier, *J. Virol.*, 2000, **74**, 10245.
154. M. Marella and J. Chabry, *J. Neurosci.*, 2004, **24**, 620.
155. L. Kercher, C. Favara, J. F. Striebel, R. LaCasse and B. Chesebro, *J. Virol.*, 2007, **81**, 10340.
156. J. Priller, M. Prinz, M. Heikenwalder, N. Zeller, P. Schwarz, F. L. Heppner and A. Aguzzi, *J. Neurosci.*, 2006, **26**, 11753.
157. P. Eikelenboom and R. Veerhuis, *Neurobiol. Aging*, 1996, **17**, 673.
158. G. J. Arlaud, C. Gaboriaud, N. M. Thielens, M. Budayova-Spano, V. Rossi and J. C. Fontecilla-Camps, *Mol. Immunol.*, 2002, **39**, 383.
159. T. Ishii, S. Haga, S. Yagishita and J. Tateishi, *Appl. Pathol.*, 1984, **2**, 370.
160. G. G. Kovacs, P. Gasque, T. Strobel, E. Lindeck-Pozza, M. Strohschneider, J. W. Ironside, H. Budka and M. Guentchev, *Neurobiol. Dis.*, 2004, **15**, 21.

161. K. Francis, J. van Beek, C. Canova, J. W. Neal and P. Gasque, *Expert Rev. Mol. Med.*, 2003, **5**, 1.
162. E. D. Walter, D. J. Stevens, A. R. Spevacek, M. P. Visconte, A. D. Rossi and G. L. Millhauser, *Curr. Protein Pept. Sci.*, 2009, **10**, 529.
163. D. R. Brown, K. Qin, J. W. Herms, A. Madlung, J. Manson, R. Strome, P. E. Fraser, T. Kruck, A. von Bohlen, W. SchulzSchaeffer, A. Giese, D. Westaway and H. Kretzschmar, *Nature*, 1997, **390**, 684.
164. P. C. Pauly and D. A. Harris, *J. Biol. Chem.*, 1998, **273**, 33107.
165. N. T. Watt and N. M. Hooper, *Trends Biochem. Sci.*, 2003, **28**, 406.
166. E. D. Walter, D. J. Stevens, M. P. Visconte and G. L. Millhauser, *J. Am. Chem. Soc.*, 2007, **129**, 15440.
167. F. Stellato, A. Spevacek, O. Proux, V. Minicozzi, G. Millhauser and S. Morante, *Eur. Biophys. J. Biophys. Lett.*, 2011, **40**, 1259.
168. E. L. Que, D. W. Domaille and C. J. Chang, *Chem. Rev.*, 2008, **108**, 1517.
169. D. R. Brown, C. Clive and S. J. Haswell, *J. Neurochem.*, 2001, **76**, 69.
170. G. Valensin, E. Molteni, D. Valensin, M. Taraszkiewicz and H. Kozlowski, *J. Phys. Chem. B*, 2009, **113**, 3277.
171. E. Aronoff-Spencer, C. S. Burns, N. I. Avdievich, G. J. Gerfen, J. Peisach, W. E. Antholine, H. L. Ball, F. E. Cohen, S. B. Prusiner and G. L. Millhauser, *Biochemistry*, 2000, **39**, 13760.
172. E. D. Walter, M. Chattopadhyay and G. L. Millhauser, *Biochemistry*, 2006, **45**, 13083.
173. H. Kozlowski, A. Janicka-Klos, P. Stanczak, D. Valensin, G. Valensin and K. Kulon, *Coord. Chem. Rev.*, 2008, **252**, 1069.
174. H. Kozlowski, M. Luczkowski and M. Remelli, *Dalton Trans.*, 2010, **39**, 6371.
175. E. Rivera-Milla, C. A. Stuermer and E. Malaga-Trillo, *Trends Genet.*, 2003, **19**, 72.
176. T. Suzuki, T. Kurokawa, H. Hashimoto and M. Sugiyama, *Biochem. Biophys. Res. Commun.*, 2002, **294**, 912.
177. F. L. Garcia, R. Zahn, R. Riek and K. Wuthrich, *Proc. Natl. Acad. Sci. U. S. A.*, 2000, **97**, 8334.
178. A. D. Gossert, S. Bonjour, D. A. Lysek, F. Fiorito and K. Wuthrich, *Proc. Natl. Acad. Sci. U. S. A.*, 2005, **102**, 646.
179. D. A. Lysek, C. Schorn, L. G. Nivon, V. Esteve-Moya, B. Christen, L. Calzolai, C. von Schroetter, F. Fiorito, T. Herrmann, P. Guntert and K. Wuthrich, *Proc. Natl. Acad. Sci. U. S. A.*, 2005, **102**, 640.
180. E. Gralka, D. Valensin, E. Porciatti, C. Gajda, E. Gaggelli, G. Valensin, W. Kamysz, R. Nadolny, R. Guerrini, D. Bacco, M. Remelli and H. Kozlowski, *Dalton Trans.*, 2008, 5207.
181. M. P. Hornshaw, J. R. McDermott and J. M. Candy, *Biochem. Biophys. Res. Commun.*, 1995, **207**, 621.
182. D. La Mendola, R. P. Bonomo, G. Impellizzeri, G. Maccarrone, G. Pappalardo, A. Pietropaolo, E. Rizzarelli and V. Zito, *J. Biol. Inorg. Chem.*, 2005, **10**, 463.

183. E. Gaggelli, E. Jankowska, H. Kozlowski, A. Marcinkowska, C. Migliorini, P. Stanczak, D. Valensin and G. Valensin, *J. Phys. Chem. B*, 2008, **112**, 15140.

184. P. Stanczak, D. Valensin, E. Porciatti, E. Jankowska, Z. Grzonka, E. Molteni, E. Gaggelli, G. Valensin and H. Kozlowski, *Biochemistry*, 2006, **45**, 12227.

185. G. Forloni, N. Angeretti, R. Chiesa, E. Monzani, M. Salmona, O. Bugiani and F. Tagliavini, *Nature*, 1993, **362**, 543.

186. R. P. Bonomo, V. Cucinotta, A. Giuffrida, G. Impellizzeri, A. Magri, G. Pappalardo, E. Rizzarelli, A. M. Santoro, G. Tabbi and L. I. Vagliasindi, *Dalton Trans.*, 2005, 150.

187. M. Luczkowski, H. Kozlowski, M. Stawikowski, K. Rolka, E. Gaggelli, D. Valensin and G. Valensin, *J. Chem. Soc., Dalton Trans.*, 2002, 2269.

188. C. Migliorini, E. Porciatti, M. Luczkowski and D. Valensin, *Coord. Chem. Rev.*, 2012, **256**, 352.

189. T. Miura, A. Horii and H. Takeuchi, *FEBS Lett.*, 1996, **396**, 248.

190. E. S. Riihimaki, J. M. Martinez and L. Kloo, *J. Phys. Chem. B*, 2007, **111**, 10529.

191. P. Stanczak, M. Luczkowski, P. Juszczyk, Z. Grzonka and H. Kozlowski, *Dalton Trans.*, 2004, 2102.

192. P. Stanczak, D. Valensin, P. Juszczyk, Z. Grzonka, C. Migliorini, E. Molteni, G. Valensin, E. Gaggelli and H. Kozlowski, *Biochemistry*, 2005, **44**, 12940.

193. C. S. Burns, E. Aronoff-Spencer, C. M. Dunham, P. Lario, N. I. Avdievich, W. E. Antholine, M. M. Olmstead, A. Vrielink, G. J. Gerfen, J. Peisach, W. G. Scott and G. L. Millhauser, *Biochemistry*, 2002, **41**, 3991.

194. T. Miura, S. Sasaki, A. Toyama and H. Takeuchi, *Biochemistry*, 2005, **44**, 8712.

195. F. H. Ruiz, E. Silva and N. C. Inestrosa, *Biochem. Biophys. Res. Commun.*, 2000, **269**, 491.

196. M. A. Wells, C. Jelinska, L. L. P. Hosszu, C. J. Craven, A. R. Clarke, J. Collinge, J. P. Waltho and G. S. Jackson, *Biochem. J.*, 2006, **400**, 501.

197. D. Valensin, M. Luczkowski, F. M. Mancini, A. Legowska, E. Gaggelli, G. Valensin, K. Rolka and H. Kozlowski, *Dalton Trans.*, 2004, 1284.

198. M. Chattopadhyay, E. D. Walter, D. J. Newell, P. J. Jackson, E. Aronoff-Spencer, J. Peisach, G. J. Gerfen, B. Bennett, W. E. Antholine and G. L. Millhauser, *J. Am. Chem. Soc.*, 2005, **127**, 12647.

199. G. L. Millhauser, *Acc. Chem. Res.*, 2004, **37**, 79.

200. W. Rachidi, D. Vilette, P. Guiraud, M. Arlotto, J. Riondel, H. Laude, S. Lehmann and A. Favier, *J. Biol. Chem.*, 2003, **278**, 9064.

201. F. Berti, E. Gaggelli, R. Guerrini, A. Janicka, H. Kozlowski, A. Legowska, H. Miecznikowska, C. Migliorini, R. Pogni, M. Remelli, K. Rolka, D. Valensin and G. Valensin, *Chem.–Eur. J.*, 2007, **13**, 1991.

202. M. Klewpatinond and J. H. Viles, *Biochem. J.*, 2007, **404**, 393.

203. K. Osz, Z. Nagy, G. Pappalardo, G. Di Natale, D. Sanna, G. Micera, E. Rizzarelli and I. Sovago, *Chem.–Eur. J.*, 2007, **13**, 7129.

204. M. Klewpatinond and J. H. Viles, *FEBS Lett.*, 2007, **581**, 1430.

205. M. Remelli, D. Valensin, D. Bacco, E. Gralka, R. Guerrini, C. Migliorini and H. Kozlowski, *New J. Chem.*, 2009, **33**, 2300.

206. L. Rivillas-Acevedo, R. Grande-Aztatzi, I. Lomeli, J. E. Garcia, E. Barrios, S. Teloxa, A. Vela and L. Quintanar, *Inorg. Chem.*, 2011, **50**, 1956.

207. C. S. Burns, E. Aronoff-Spencer, G. Legname, S. B. Prusiner, W. E. Antholine, G. J. Gerfen, J. Peisach and G. L. Millhauser, *Biochemistry*, 2003, **42**, 6794.

208. E. Gralka, D. Valensin, M. Remelli and H. Kozlowski, in *Brain Diseases and Metalloproteins*, ed. D. R. Brown, Pan Stanford, Singapore, 2012, p. 33.

209. O. Yamauchi, A. Odani and M. Takani, *J. Chem. Soc., Dalton Trans.*, 2002, 3411.

210. G. Di Natale, G. Grasso, G. Impellizzeri, D. La Mendola, G. Micera, N. Mihala, Z. Nagy, K. Osz, G. Pappalardo, V. Rigo, E. Rizzarelli, D. Sanna and I. Sovago, *Inorg. Chem.*, 2005, **44**, 7214.

211. J. Shearer, P. Soh and S. Lentz, *J. Inorg. Biochem.*, 2008, **102**, 2103.

212. D. Valensin, K. Gajda, E. Gralka, G. Valensin, W. Kamysz and H. Kozlowski, *J. Inorg. Biochem.*, 2010, **104**, 71.

213. E. Gralka, D. Valensin, K. Gajda, D. Bacco, L. Szyrwiel, M. Remelli, G. Valensin, W. Kamasz, W. Baranska-Rybak and H. Kozlowski, *Mol. Biosys.*, 2009, **5**, 497.

214. P. Davies and D. R. Brown, *Biochem. J.*, 2008, **410**, 237.

215. A. Rana, D. Gnaneswari, S. Bansal and B. Kundu, *Chem.-Biol. Interact.*, 2009, **181**, 282.

216. M. P. Hornshaw, J. R. McDermott, J. M. Candy and J. H. Lakey, *Biochem. Biophys. Res. Commun.*, 1995, **214**, 993.

217. J. H. Viles, F. E. Cohen, S. B. Prusiner, D. B. Goodin, P. E. Wright and H. J. Dyson, *Proc. Natl. Acad. Sci. U. S. A.*, 1999, **96**, 2042.

218. R. M. Whittal, H. L. Ball, F. E. Cohen, A. L. Burlingame, S. B. Prusiner and M. A. Baldwin, *Protein Sci.*, 2000, **9**, 332.

219. G. S. Jackson, I. Murray, L. L. P. Hosszu, N. Gibbs, J. P. Waltho, A. R. Clarke and J. Collinge, *Proc. Natl. Acad. Sci. U. S. A.*, 2001, **98**, 8531.

220. A. P. Garnett and J. H. Viles, *J. Biol. Chem.*, 2003, **278**, 6795.

221. G. Arena, D. La Mendola, G. Pappalardo, I. Sovago and E. Rizzarelli, *Coord. Chem. Rev.*, 2012, **256**, 2202.

222. A. G. Kenward, L. J. Bartolotti and C. S. Burns, *Biochemistry*, 2007, **46**, 4261.

223. D. R. Brown, F. Hafiz, L. L. Glasssmith, B. S. Wong, I. M. Jones, C. Clive and S. J. Haswell, *EMBO J.*, 2000, **19**, 1180.

224. M. Klewpatinond, P. Davies, S. Bowen, D. R. Brown and J. H. Viles, *J. Biol. Chem.*, 2008, **283**, 1870.

225. K. M. Pan, N. Stahl and S. B. Prusiner, *Protein Sci.*, 1992, **1**, 1343.

226. R. S. Mishra, S. Basu, Y. P. Gu, X. Luo, W. Q. Zou, R. Mishra, R. L. Li, S. G. Chen, P. Gambetti, H. Fujioka and N. Singh, *J. Neurosci.*, 2004, **24**, 11280.

227. S. Fernaeus, J. Halldin, L. Bedecs and T. Land, *Mol. Brain Res.*, 2005, **133**, 266.

228. S. Basu, M. L. Mohan, X. Luo, B. Kundu, Q. Z. Kong and N. Singh, *Mol. Biol. Cell*, 2007, **18**, 3302.

229. D. R. Brown, *Metallomics*, 2011, **3**, 229.

230. H. Kozlowski, M. Luczkowski, M. Remelli and D. Valensin, *Coord. Chem. Rev.*, 2012, **256**, 2129.

231. M. W. Brazier, P. Davies, E. Player, F. Marken, J. H. Viles and D. R. Brown, *J. Biol. Chem.*, 2008, **283**, 12831.

232. A. M. Thackray, R. Knight, S. J. Haswell, R. Bujdoso and D. R. Brown, *Biochem. J.*, 2002, **362**, 253.

233. B. S. Wong, S. G. Chen, M. Colucci, Z. L. Xie, T. Pan, T. Liu, R. L. Li, P. Gambetti, M. S. Sy and D. R. Brown, *J. Neurochem.*, 2001, **78**, 1400.

234. S. Hesketh, J. Sassoon, R. Knight and D. R. Brown, *Mol. Cell. Neurosci.*, 2008, **37**, 590.

235. S. Hesketh, J. Sassoon, R. Knight, J. Hopkins and D. R. Brown, *J. Anim. Sci.*, 2007, **85**, 1596.

236. F. Zhu, P. Davies, A. R. Thompsett, S. M. Kelly, G. E. Tranter, L. Hecht, N. W. Isaacs, D. R. Brown and L. D. Barron, *Biochemistry*, 2008, **47**, 2510.

237. C. J. Choi, V. Anantharam, N. J. Saetveit, R. S. Houk, A. Kanthasamy and A. G. Kanthasamy, *Toxicol. Sci.*, 2007, **98**, 495.

238. E. Gaggelli, F. Bernardi, E. Molteni, R. Pogni, D. Valensin, G. Valensin, M. Remelli, M. Luczkowski and H. Kozlowski, *J. Am. Chem. Soc.*, 2005, **127**, 996.

239. M. D. Garrick, S. T. Singleton, F. Vargas, H. C. Kuo, L. Zhao, M. Knopfel, T. Davidson, M. Costa, P. Paradkar, J. A. Roth and L. M. Garrick, *Biol. Res.*, 2006, **39**, 79.

240. N. D. Younan, M. Klewpatinond, P. Davies, A. V. Ruban, D. R. Brown and J. H. Viles, *J. Mol. Biol.*, 2011, **410**, 369.

241. O. V. Bocharova, L. Breydo, V. V. Salnikov and I. V. Baskakov, *Biochemistry*, 2005, **44**, 6776.

242. N. Hijazi, Y. Shaked, H. Rosenmann, T. Ben-Hur and R. Gabizon, *Brain Res.*, 2003, **993**, 192.

243. M. Pushie, A. Rauk, F. Jirik and H. Vogel, *Biometals*, 2009, **22**, 159.

244. M. J. Pushie and H. J. Vogel, *J. Toxicol. Environ. Health*, 2009, **72**, 1040.

245. E. M. Sigurdsson, D. R. Brown, M. A. Alim, H. Scholtzova, R. Carp, H. C. Meeker, F. Prelli, B. Frangione and T. Wisniewski, *J. Biol. Chem.*, 2003, **278**, 46199.

246. S. Morante, R. Gonzalez-Iglesias, C. Potrich, C. Meneghini, W. Meyer-Klaucke, G. Menestrina and M. Gasset, *J. Biol. Chem.*, 2004, **279**, 11753.

247. R. N. Tsenkova, I. K. Iordanova, K. Toyoda and D. R. Brown, *Biochem. Biophys. Res. Commun.*, 2004, **325**, 1005.

248. N. H. Kim, J. K. Choi, B. H. Jeong, J. I. Kim, M. S. Kwon, R. I. Carp and Y. S. Kim, *FASEB J.*, 2005, **19**, 783.

249. P. Davies, F. Marken, S. Salter and D. R. Brown, *Biochemistry*, 2009, **48**, 2610.

250. A. Giese, M. Buchholz, J. Herms and H. A. Kretzschmar, *J. Mol. Neurosci.*, 2005, **27**, 347.

251. S. Fernaeus and T. Land, *Neurosci. Lett.*, 2005, **382**, 217.

252. W. Sumudhu, S. Perera and N. M. Hooper, *Curr. Biol.*, 2001, **11**, 519.

253. J. A. Tainer, E. D. Getzoff, K. M. Beem, J. S. Richardson and D. C. Richardson, *J. Mol. Biol.*, 1982, **160**, 181.

254. D. R. Brown and A. Besinger, *Biochem. J.*, 1998, **334**, 423.

255. S. Jones, M. Batchelor, D. Bhelt, A. R. Clarke, J. Collinge and G. S. Jackson, *Biochem. J.*, 2005, **392**, 309.

256. G. Hutter, F. L. Heppner and A. Aguzzi, *Biol. Chem.*, 2003, **384**, 1279.

257. B. Boka, A. Myari, I. Sovago and N. Hadjiliadis, *J. Inorg. Biochem.*, 2004, **98**, 113.

258. A. Jancso, Z. Paksi, N. Jakab, B. Gyurcsik, A. Rockenbauer and T. Gajda, *Dalton Trans.*, 2005, 3187.

259. D. La Mendola, R. P. Bonomo, S. Caminati, G. Di Natale, S. S. Emmi, O. Hansson, G. Maccarrone, G. Pappalardo, A. Pietropaolo and E. Rizzarelli, *J. Inorg. Biochem.*, 2009, **103**, 195.

260. P. Stanczak and H. Kozlowski, *Biochem. Biophys. Res. Commun.*, 2007, **352**, 198.

261. H. Ohtsu, Y. Shimazaki, A. Odani, O. Yamauchi, W. Mori, S. Itoh and S. Fukuzumi, *J. Am. Chem. Soc.*, 2000, **122**, 5733.

262. A. M. Haeberle, C. Ribaut-Barassin, G. Bombarde, J. Mariani, G. Hunsmann, J. Grassi and Y. Bailly, *Microsc. Res. Tech.*, 2000, **50**, 66.

263. J. Herms, T. Tings, S. Gall, A. Madlung, A. Giese, H. Siebert, P. Schurmann, O. Windl, N. Brose and H. Kretzschmar, *J. Neurosci.*, 1999, **19**, 8866.

264. N. Sales, K. Rodolfo, R. Hassig, B. Faucheux, L. Di Giamberardino and K. L. Moya, *Eur. J. Neurosci.*, 1998, **10**, 2464.

265. A. J. Cross, R. H. Kimberlin, T. J. Crow, J. A. Johnson and C. A. Walker, *J. Neurol. Sci.*, 1985, **70**, 231.

266. H. G. Lee, S. J. Park, E. K. Choi, R. I. Carp and Y. S. Kim, *J. Mol. Neurosci.*, 1999, **13**, 121.

267. L. Re, F. Rossini, F. Re, M. Bordicchia, A. Mercanti, O. S. L. Fernandez and S. Barocci, *Pharmacol. Res.*, 2006, **53**, 62.

268. R. Linden, V. R. Martins, M. A. M. Prado, M. Cammarota, I. Izquierdo and R. R. Brentani, *Physiol. Rev.*, 2008, **88**, 673.

269. L. Varela-Nallar, E. M. Toledo, L. F. Larrondo, A. L. B. Cabral, V. R. Martins and N. C. Inestrosa, *Am. J. Physiol.: Cell Physiol.*, 2006, **290**, C271.

270. J. Collinge, M. A. Whittington, K. C. L. Sidle, C. J. Smith, M. S. Palmer, A. R. Clarke and J. G. R. Jefferys, *Nature*, 1994, **370**, 295.

271. A. Rangel, N. Madronal, A. Gruart, R. Gavin, F. Llorens, L. Sumoy, J. M. Torres, J. M. Delgado-Garcia and J. A. Del Rio, *PLoS One*, 2009, **4**, e7592.
272. H. Diringer and B. Ehlers, *J. Gen. Virol.*, 1991, **72**, 457.
273. B. Ehlers and H. Diringer, *J. Gen. Virol.*, 1984, **65**, 1325.
274. L. Ingrosso, A. Ladogana and M. Pocchiari, *J. Virol.*, 1995, **69**, 506.
275. S. A. Priola, A. Raines and W. S. Caughey, *Science*, 2000, **287**, 1503.
276. C. Korth, B. C. H. May, F. E. Cohen and S. B. Prusiner, *Proc. Natl. Acad. Sci. U. S. A.*, 2001, **98**, 9836.
277. J. Delay, P. Deniker and J. M. Harl, *Ann. Med. Psychol. (Paris)*, 1952, **110**, 267.
278. L. S. Goodman and A. Gilman, *The Pharmacological Basis of Therapeutics*, Macmillan, New York, 1975.
279. J. Collinge, M. Gorham, F. Hudson, A. Kennedy, G. Keogh, S. Pal, M. Rossor, P. Rudge, D. Siddique, M. Spyer, D. Thomas, S. Walker, T. Webb, S. Wroe and J. Darbyshire, *Lancet Neurol.*, 2009, **8**, 334.
280. C. F. Farquhar and A. G. Dickinson, *J. Gen. Virol.*, 1986, **67**, 463.
281. R. H. Kimberlin and C. A. Walker, *Antimicrob. Agents Chemother.*, 1986, **30**, 409.
282. B. Caughey and G. J. Raymond, *J. Virol.*, 1993, **67**, 643.
283. C. Farquhar, A. Dickinson and M. Bruce, *Lancet*, 1999, **353**, 117.
284. R. H. Kimberlin and C. A. Walker, *Arch. Virol.*, 1983, **78**, 9.
285. A. Ladogana, P. Casaccia, L. Ingrosso, M. Cibati, M. Salvatore, Y. G. Xi, C. Masullo and M. Pocchiari, *J. Gen. Virol.*, 1992, **73**, 661.
286. I. R. Macgregor, J. Dawes, L. Paton, D. S. Pepper, C. V. Prowse and M. Smith, *Thromb. Haemost.*, 1984, **51**, 321.
287. K. Doh-Ura, K. Ishikawa, I. Murakami-Kubo, K. Sasaki, S. Mohri, R. Race and T. Iwaki, *J. Virol.*, 2004, **78**, 4999.
288. I. Bone, L. Belton, A. S. Walker and J. Darbyshire, *Eur. J. Neurol.*, 2008, **15**, 458.
289. Y. Tsuboi, K. Doh-ura and T. Yamada, *Neuropathology*, 2009, **29**, 632.
290. N. G. Rainov, Y. Tsuboi, P. Krolak-Salmon, A. Vighetto and K. Doh-ura, *Expert Opin. Biol. Ther.*, 2007, **7**, 713.
291. N. V. Todd, J. Morrow, K. Doh-ura, S. Dealler, S. O'Hare, P. Farling, M. Duddy and N. G. Rainov, *J. Infect.*, 2005, **50**, 394.
292. M. Otto, L. Cepek, P. Ratzka, S. Doehlinger, I. Boekhoff, J. Wiltfang, E. Irle, G. Pergande, B. Ellers-Lenz, O. Windl, H. A. Kretzschmar, S. Poser and H. Prange, *Neurology*, 2004, **62**, 714.
293. C. Rodriguez-Rodriguez, M. Telpoukhovskaia and C. Orvig, *Coord. Chem. Rev.*, 2012, **256**, 2308.
294. N. Singh, D. Das, A. Singh and M. L. Mohan, *Curr. Issues Mol. Biol.*, 2010, **12**, 99.
295. M. Stefani and C. M. Dobson, *J. Mol. Med.*, 2003, **81**, 678.
296. M. L. Bolognesi, A. Cavalli, C. Bergarmini, R. Fato, G. Lenaz, M. Rosini, M. Bartolini, V. Andrisano and C. Melchiorre, *J. Med. Chem.*, 2009, **52**, 7883.

297. J. J. Braymer, J. S. Choi, A. S. DeToma, C. Wang, K. Nam, J. W. Kampf, A. Ramamoorthy and M. H. Lim, *Inorg. Chem.*, 2011, **50**, 10724.

298. H. X. Jin, J. Randazzo, P. Zhang and P. F. Kador, *J. Med. Chem.*, 2010, **53**, 1117.

299. S. Mandel, T. Amit, O. Bar-Am and M. B. H. Youdim, *Prog. Neurobiol.*, 2007, **82**, 348.

300. B. Meunier, *Acc. Chem. Res.*, 2008, **41**, 69.

301. H. Schugar, D. E. Green, M. L. Bowen, L. E. Scott, T. Storr, K. Bohmerle, F. Thomas, D. D. Allen, P. R. Lockman, M. Merkel, K. H. Thompson and C. Orvig, *Angew. Chem. Int. Ed.*, 2007, **46**, 1716.

302. M. B. Youdim, *Curr. Alzheimer Res.*, 2006, **3**, 541.

303. H. L. Zheng, L. M. Weiner, O. Bar-Am, S. Epsztejn, Z. I. Cabantchik, A. Warshawsky, N. B. H. Youdim and M. Fridkin, *Bioorg. Med. Chem.*, 2005, **13**, 773.

304. J. A. Duce and A. I. Bush, *Prog. Neurobiol.*, 2010, **92**, 1.

305. R. A. Cherny, J. T. Legg, C. A. McLean, D. P. Fairlie, X. D. Huang, C. S. Atwood, K. Beyreuther, R. E. Tanzi, C. L. Masters and A. I. Bush, *J. Biol. Chem.*, 1999, **274**, 23223.

306. A. I. Bush, *Trends Neurosci.*, 2003, **26**, 207.

307. C. W. Ritchie, A. I. Bush, A. Mackinnon, S. Macfarlane, M. Mastwyk, L. MacGregor, L. Kiers, R. Cherny, Q. X. Li, A. Tammer, D. Carrington, C. Mavros, I. Volitakis, M. Xilinas, D. Ames, S. Davis, I. Volitakis, M. Xilinas, D. Ames, S. Davis, K. Beyreuther, R. E. Tanzi and C. L. Masters, *Arch. Neurol.*, 2003, **60**, 1685.

Multiple Sclerosis

RICHARD REYNOLDS AND ROBERT CRICHTON*

Université Catholique de Louvain, Belgium
*Email: robert.crichton@uclouvain.be

7.1 Introduction

Neurodegenerative diseases can often affect cognitive function [*e.g.* mild cognitive impairment (MCI), Alzheimer's disease], motor function (Parkinson's disease) or various combinations of both. In this chapter we describe multiple sclerosis (MS), the prototype inflammatory autoimmune disorder of the central nervous system (CNS),[1] which causes demyelination of the CNS, resulting in progressive loss of motor and sensory function. MS is the most prevalent demyelinating disease, is one of the most common reasons for admission to a neurological hospital ward and represents the principal cause of neurological disability in young adults, with a prevalence of approximately 120 per 100 000 and a lifetime risk of 1 in 400.[2] Although MS has been studied for over 150 years, we are still unsure what mechanisms underlie the generation of the characteristic inflammatory demyelinating lesions in the CNS, and how they give rise to both acute and chronic progressive neurological dysfunction.

There are approximately 2.5 million individuals throughout the world affected by MS, and the disease affects northern Europeans (Figure 7.1) with a much greater frequency than other populations.[2,3] The highest incidence is in Scotland (population 5 million), with over 10 500 cases compared to 70 000 for the UK as a whole (population 60 million). Other high frequency areas comprise most of northern Europe, Israel, Canada, northern US, southeastern Australia, New Zealand and easternmost Russia. However, the distribution of MS is not due to population genetics alone, since the migration of immigrants

RSC Metallobiology Series No. 1
Mechanisms and Metal Involvement in Neurodegenerative Diseases
Edited by Roberta Ward, David Dexter and Robert Crichton
© The Royal Society of Chemistry 2013
Published by the Royal Society of Chemistry, www.rsc.org

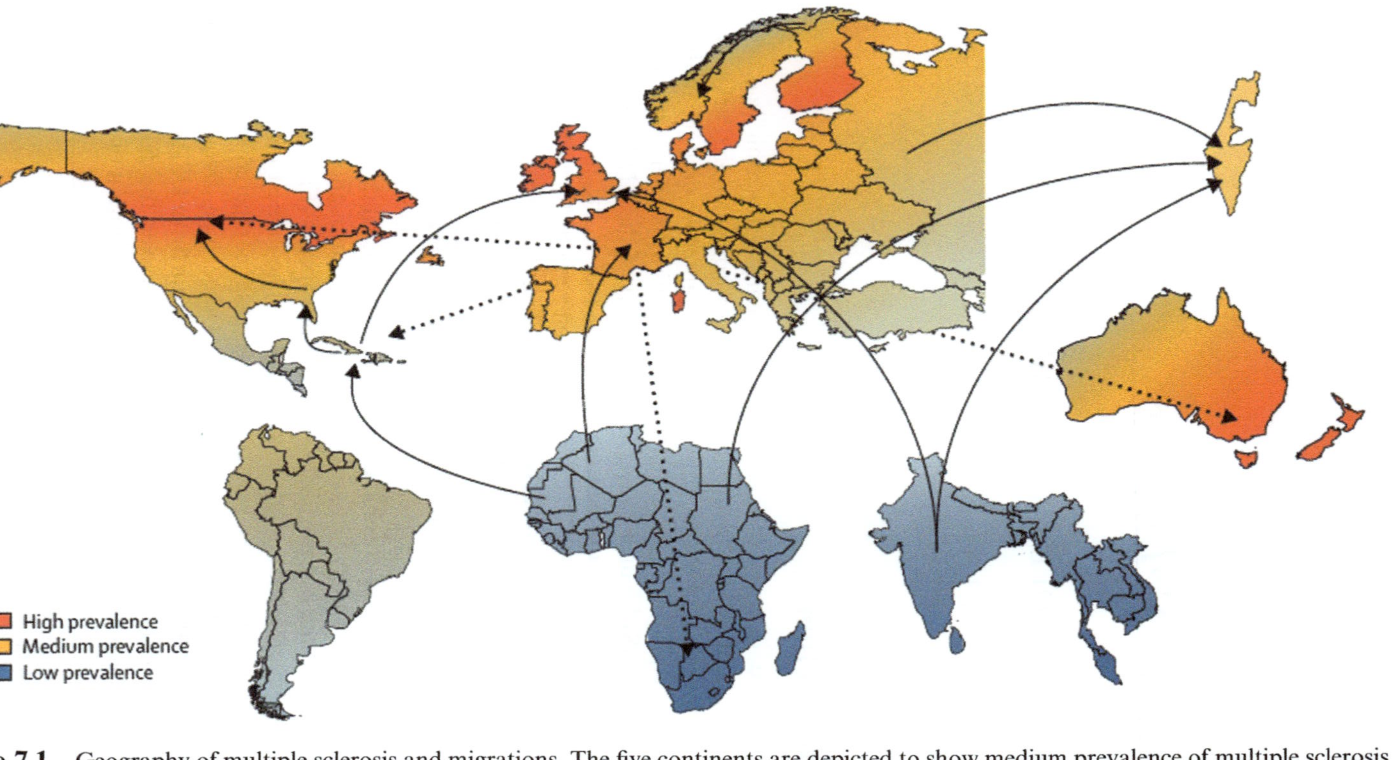

Figure 7.1 Geography of multiple sclerosis and migrations. The five continents are depicted to show medium prevalence of multiple sclerosis (*orange*), areas of exceptionally high frequency (*red*) and those with low rates (*grey-blue*). Some regions are fairly uncharted and these colours are only intended to provide an impression of the geographical trends. Major routes of migration from the high-risk zone of northern Europe, especially including small but informative studies, are shown as *dotted arrows*. Studies involving migrants from low-risk to high-risk zones are shown as *solid arrows*.
(Reproduced from Compston and Coles[3] with permission from Elsevier).

also affects the distribution, and many studies correlate the risk of multiple sclerosis with the place of residence in childhood.[2] In common with other autoimmune diseases, MS affects twice as many women as it does men.

7.2 Definition of the Disease and Clinical Presentation

MS is an autoimmune disorder of the CNS characterized by inflammatory destruction of the myelin sheaths of axons in the CNS. Myelin forms a sort of "electrical insulating tape", which is wrapped around the axon (Figure 7.2a), electrically isolating it from the surrounding medium. Myelin is produced by specialized cells: the oligodendrocytes in the CNS and the Schwann cells in the peripheral nervous system. The myelin sheaths, which are in fact the plasma

(a)

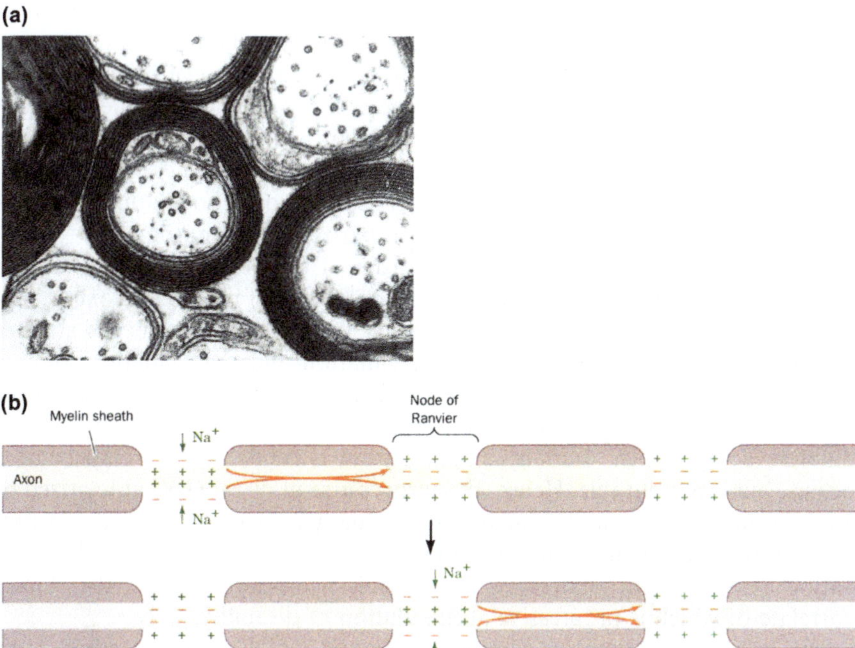

Figure 7.2 (a) An electron micrograph of myelinated nerve fibers in cross section. The myelin sheath surrounding the axon is the plasma membrane of an oligodendrocyte or a Schwann cell, which extrudes its cytoplasm from between the layers as it grows spirally around the axon. The resulting bilayer, which makes between 10 and 150 turns about the axon, is a good conductor of electricity because of its high lipid content (about 80%). (b) A schematic diagram of a myelinated axon in longitudinal cross section. The axonal membrane is in contact with the external medium at the nodes of Ranvier. A depolarization generated by an action potential at one node hops down the myelinated axon (*arrows*) to the neighbouring node, where it induces a new action potential. This results in saltatory conduction. (Reproduced from Voet and Voet[107] with permission from Wiley).

membrane of the oligodendrocyte or Schwann cell, wrap themselves around the axons. The resulting bilayer, which makes between 10 and 50 turns around the axon, is a good electrical insulator because of its high lipid content (about 80%). Nerve impulses in myelinated nerves travel much faster than in non-myelinated nerves, up to 100 m s^{-1} compared to a maximum of 10 m s^{-1}. This is because the myelinated axonal membrane is in contact with the external medium only at the myelin-free nodes of Ranvier axons, where the Na$^+$ channels are uniquely localized, compared to their rather sparse distribution in non-myelinated axons (Figure 7.2b). A depolarization generated by an action potential at one node literally hops down the myelinated axon to the neighbouring node of Ranvier, where it induces a new action potential. This process is termed saltatory conduction.

The oligodendrocyte is a principal target of attack in MS, and the consequences of demyelination for saltatory conduction explain many of the features of the disease. Partially demyelinated axons conduct nerve impulses more slowly. Demyelinated axons can discharge spontaneously and show increased mechanical sensitivity, accounting for flashes of light on eye movement and electrical sensations running down the spine or limbs on flexing the neck. Axons which have lost part of their myelin sheath are unable to sustain the fall in membrane capacitance induced by a rise in temperature, and conduction fails, leading to appearance of symptoms after exercise or a hot bath. Cross-talk between demyelinated axons may result in paroxysmal symptoms.

MS is primarily an inflammatory disorder of the brain and spinal cord in which focal lymphocytic infiltration leads to damage of myelin and axons. The hallmark is formation of the sclerotic plaque, which represents the end stage of a process involving inflammation, demyelination and remyelination, oligodendrocyte depletion and astrocytosis, and neuronal and axon degeneration, although there is no consensus concerning the order and relation of these different components.[3] Initially, inflammation is transient and remyelination occurs, but is not durable. Hence, the early course of disease is characterized by episodes of neurological dysfunction that usually recover. However, over time the pathological changes become dominated by widespread microglial activation associated with extensive and chronic neurodegeneration, the clinical correlate of which is progressive accumulation of disability.

Diagnosis of MS involves clinical and paraclinical diagnostic methods integrated with magnetic resonance imaging (MRI), using currently accepted diagnostic criteria.[4] MRI shows abnormalities that indicate foci of blood–brain barrier (BBB) breakdown and demyelination of CNS white matter; visually evoked electrophysiological potentials reveal interference of conduction in previously myelinated pathways; and examination of the cerebrospinal fluid indicates the intrathecal synthesis of oligoclonal antibodies.

Although the exact mechanisms that are involved in the pathogenesis of MS remain unknown, it has been proposed that lymphocytes activated in the periphery migrate to the CNS, where they become attached to receptors on endothelial cells of post-capillary venules, and then cross the BBB. T cells are then reactivated by fragments of myelin antigens exposed on the surface of antigen presenting cells (macrophages and microglia). This reactivation induces the release of proinflammatory cytokines which further open up the BBB, resulting in a second, larger wave of inflammatory cell recruitment into the CNS.[5]

Susceptibility to development of MS appears to be associated with multiple factors, and a large corpus of data indicates that immune processes play an essential role in driving the disease process.[6] Major support for an autoimmune hypothesis comes from studies of experimental autoimmune encephalomyelitis (EAE), an experimental disease in animals, which can be induced by active immunization with myelin antigens in genetically susceptible strains.[7] EAE shares some clinical and pathological features with MS,[8] and is generally considered by immunologists to be an animal model of the early stages of MS.[9]

7.3   Genetic Contribution

Multiple sclerosis is thought to be triggered and driven by environmental and genetic/epigenetic factors. Established environmental factors include lower exposure to UV radiation, lower vitamin D levels and the Epstein–Barr virus.[10] Epidemiological studies established that genetic factors are primarily responsible for the increased frequency of the disease seen in the relatives of affected individuals.[11,12] Systematic attempts to identify linkage in multiplex families have confirmed, as was already shown nearly 40 years ago, that variation within the major histocompatibility complex (MHC) exerts the greatest individual effect on risk.[13] Advances in genome technology have enabled the effective screening of thousands of single nucleotide polymorphisms (SNPs) in thousands of samples at an affordable price, leading to the advent of genome-wide association studies (GWAS). In contrast to earlier candidate gene studies, the strategy of GWAS is free of any preconceived hypotheses and aims to screen the whole genome, and has permitted the identification of additional risk variants. A series of GWAS studies and meta-analyses have been conducted in different MS cohorts, which, by the summer of 2011, together with follow-up studies, had confirmed 26 loci with genome-wide significance.[14,15] Many of these common risk variants are located at or near genes with central immunological functions and the majority are associated with other autoimmune diseases. However, most of the genetic architecture underlying susceptibility to the disease remains to be defined and is anticipated to require the analysis of sample sizes that are beyond the numbers currently available to individual research groups. A collaborative GWAS involving 9772 cases of European descent and 17 376 controls, collected by 23 research groups working in 15 different countries, has been recently published, which has replicated almost all of the previously suggested associations and identified at least a further 29 novel susceptibility loci.[16] Immunologically relevant genes are significantly overrepresented among those mapping close to the identified loci and particularly implicate T-helper-cell differentiation in the pathogenesis of multiple sclerosis (Figure 7.3). However, the mechanisms by which all of the identified genes affect susceptibility is still unknown.

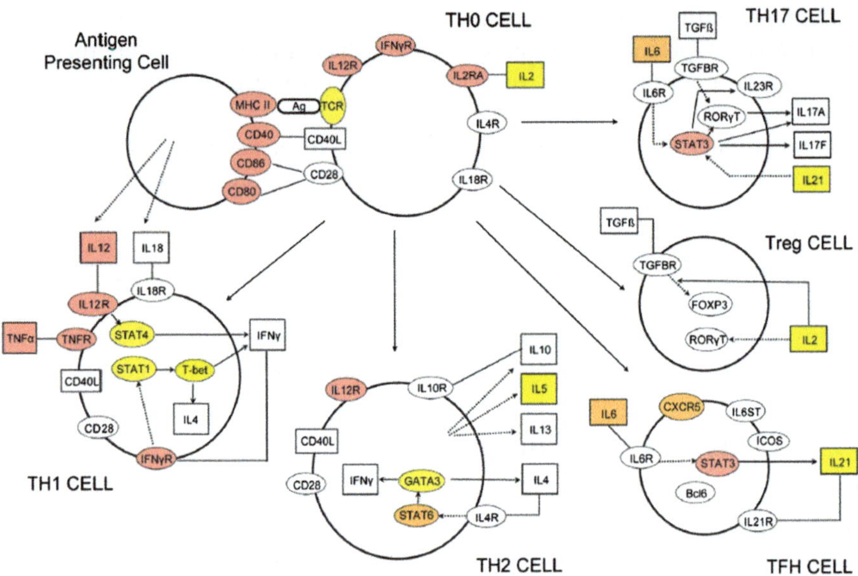

Figure 7.3   Graphical representation of the T-cell differentiation pathway. Alpha-numeric labels indicate the individual genes and gene complexes (nodes) included in the pathway (some are labelled more than once). Coloured nodes are those containing a gene implicated by proximity to an SNP showing evidence of association with MS. In decreasing frequency: *red, orange, yellow*. Reproduced from Du and Xie[17] with permission from Nature Publishing Group.

7.4   Proteins Possibly Involved in the Disease and Associated Pathologies

It should be clear from the preceding section that the more than 50 proteins encoded by the genes implicated in the collaborative GWAS,[16] reported above, could have a possible involvement in MS. However, it needs to be remembered that the GWAS studies have only identified genes that contribute to suscepti-bility to develop MS, and not to the subsequent clinical disease course. We present some of these in what follows in no particular order.

Within the MHC the collaborative GWAS study confirmed the previous associations, refined the identity of the *DRB1* risk alleles and confirmed that variation in the *HLA-A* gene underlies the protective effect which had been attributed to the Class 1 region.[16]

The award of the 2012 Nobel Prize in Chemistry to Robert Lefkovich and Brian Kobilka for "their ground-breaking discoveries that reveal the inner workings of … G-protein-coupled receptors" underlines the key role which G-protein-coupled receptors (GPCRs) play in enabling the cells of our body to sense their environment, mediating our physiological responses to hormones, neurotransmitters and environmental stimulants. It therefore comes as no surprise that they are considered as important therapeutic targets for a broad spectrum of diseases. A considerable body of data from animal and clinical studies indicates that many GPCRs are critically involved in various aspects of MS pathogenesis, including antigen presentation, cytokine production, T-cell differentiation, T-cell proliferation, T-cell invasion, *etc.*[17] As we will see later, sphingosine 1-phosphate (S1P) receptors, a subset of a larger family of cell-surface GPCRs which mediate the effects of lysophospholipids, are the target of the orally active MS drug fingolimod.[18]

The role of ABC transporters in infectious and autoimmune inflammatory diseases of the CNS has gained considerable attention, and in MS and its animal model EAE, ABC transporters may be involved both in pathogenesis and response to treatment. Genetic variations in ABC transporters appear to be responsible for the variable response to treatment and potentially adverse fatal side-effect events of mitoxantrone, a therapeutic agent used in aggressive MS.[19]

Toll-like receptors (TLR) are proteins of the innate immune system involved in the identification and clearance of invading pathogens, which transmit their signal through adaptor proteins, most commonly myeloid differentiation primary response gene 88 (MyD88). Inappropriate response of specific TLR has been implicated in a number of autoimmune diseases, notably MS. Activation of TLR2, TLR4, TLR7 and TLR9 are thought to play a role in EAE, a murine model of MS, whereas activation of TLR3 protects from disease.[20]

Semaphorins were originally identified as guidance cues in neural development. However, a number of semaphorins called "immune semaphorins", such as Sema3A, 4A, 4D, 6D and 7A, seem to be critically involved in various phases of the immune response by regulating immune cell–cell contacts or cell migration. These semaphorins and their receptors in the immune system may play an important role in the pathogenesis of MS and its animal model, EAE.[21]

Osteopontin (OPN) is a pleiotropic cytokine with multiple immunological functions which is secreted by activated macrophages, leukocytes and activated T lymphocytes, and its expression is up-regulated during inflammation. The secreted form of OPN is a T helper type 1 (Th1) cytokine and is a chemoattractant for many types of cells through integrin receptors and CD44, whereas its intracellular form is a critical regulator for Toll-like receptor-9, TLR-7-dependent interferon-alpha expression and Th17 development. Elevated levels of OPN transcripts are found in MS brain lesions and spinal cords of rats with EAE, and OPN knockout mice are protected from severe EAE. Increased levels of OPN were reported in plasma and CSF of MS patients in comparison to healthy controls. However, the involvement of OPN, in part *via* non-immune effects, in remyelination and its neuroprotective potential need to balanced against its proinflammatory properties.[22,23]

Over the last few years, evidence from MS and experimental models of neuroinflammation has suggested that autoreactive T cells could exert neuroprotective effects through the release of neurotrophins, in particular brain-derived neurotrophic factor (BDNF), by autoreactive T cells, thereby facilitating brain tissue repair. Support for this hypothesis comes from recent studies showing that glatiramer acetate, a currently approved treatment for MS, promotes the expansion of T cell clones crossing the BBB and releasing

BDNF *in situ*. A small subset of autoreactive T cells expresses the high-affinity full-length receptor for BDNF (TrkB-TK) in the periphery. In MS patients, T cells show reduced susceptibility to activation-induced apoptosis, a crucial mechanism eliminating autoreactive T clones and contributing to peripheral immunologic tolerance. These findings suggest the existence of a dual effect exerted by BDNF, which not only provides neuroprotection in the CNS could also be proinflammatory by promoting the survival of autoreactive T cells through an autocrine/paracrine loop.[24]

Tumour necrosis factor (TNF) is a cytokine with pleiotropic actions in immunity, inflammation, control of cell proliferation, differentiation and necrosis, which is involved in the pathogenesis of MS.[25] It is synthesized as a type 2 transmembrane precursor protein (tmTNF), which is then cleaved, releasing a soluble form (sTNF) into the circulation. Both sTNF and tmTNF are produced by a wide range of immune cells, particularly activated macrophages, but also T and B lymphocytes, natural killer cells, dendritic cells and monocytes. Within the CNS they are produced by microglia, astrocytes and certain neuron populations.[26,27] Both tmTNF and sTNF are biologically active in the form of homotrimers, but their activities are in direct opposition to each other. TNF signals *via* two structurally related but functionally distinct TNF receptors (TNFR1 and TNFR2) to which tmTNF and sTNF can bind (Figure 7.4). sTNF shows greater affinity for TNFR1 (dissociation constant $[K_d] \approx 20$ pM) than for TNFR2 ($K_d \approx 400$ pM). TNFR2 in turn is preferentially activated by tmTNF. sTNF is the primary mediator of TNF inflammatory responses, leading to apoptosis *via* the extrinsic pathway. TNFR2 mediated signalling leads to activation of NF-κB, which promotes cell survival. Studies in EAE showed that TNF is clearly proinflammatory during the acute phases of the disease, involving activation of peripheral immune cells and demyelination is more extensive when TNF is present. This suggested that TNF blocking agents should have a beneficial effect in MS disease activity, yet when anti-TNF agents were administered to MS patients the opposite effect was observed in some cases, with increased clinical and radiological disease activity in these patients. This is consistent with the divergent roles existing between the two receptors that mediate the TNF actions,[25] the anti-TNF agents inhibiting both signalling pathways. A single nucleotide polymorphism (SNP), rs1800693 in the

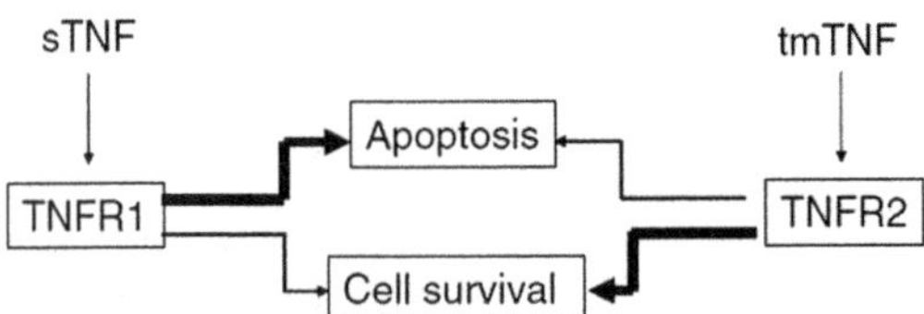

Figure 7.4   TNFR-mediated differential biological responses of TNF-α. Soluble TNF (sTNF) and transmembrane TNF (tmTNF) can bind to both receptors, TNFR1 and TNFR2, though they show certain preferences.
(Reproduced from Beutler and Cerami[27] with permission from Annual Reviews).

gene which encodes tumour necrosis factor receptor 1 (TNFR1), was found to be associated with MS, but not with other autoimmune conditions.[28] Genetic evidence strongly implicates rs1800693 as the causal variant in the TNFR1 gene and functional studies have shown that the MS risk allele directs expression of a novel, soluble form of TNFR1 which can block TNF. TNF-blocking drugs can promote onset or exacerbation of MS, but they have proven highly efficacious in the treatment of autoimmune diseases for which there is no association with rs1800693.

7.5 Neuroinflammation

The generally accepted model of MS involves an initial immune-mediated attack on myelin antigens in the CNS, in which an inflammatory response results in direct damage to oligodendrocytes and myelin and indirect damage to axons, leading to formation of acute plaques in the white matter.[29,30] These acute inflammatory events are thought to underlie the clinical relapses that characterize the early stages of MS, the anatomical site of the inflammation determining the neurological dysfunction. How the acute inflammatory lesions evolve into chronic lesions with little ongoing inflammation is less clear.

Elucidation of the inflammatory stage of the pathogenesis of MS has relied heavily on the EAE animal model, which has been very successful in reproducing the early component. Studies in this animal model have provided convincing evidence that T cells specific for self-antigens can mediate inflammatory pathology similar to that seen in MS. T helper type 1 (Th1) cells were thought to be the main effector T cells responsible for the autoimmune inflammation. However, an important pathogenic role has emerged recently for $CD4^+$ T cells that secrete interleukin (IL)-17, termed Th17, but also IL-17-secreting $\gamma\delta$ T cells and IL-9n produced by a specific subset of T helper cells (Th9 cells) in EAE and other autoimmune and chronic inflammatory conditions.[31,32] Early clinical trials of monoclonal antibodies against IL-17 or its receptor (IL-17R) have given promising results in patients with MS.

The migration and effector function of T cells in the CNS in EAE is illustrated in Figure 7.5.[31] After immunization with myelin antigens, dendritic cells (DCs) are activated in the lymph nodes by TLR agonists and present myelin antigen to naive T cells. The activated myelin-specific T cells enter the bloodstream and cross into the CNS. Transient breakdown of the BBB occurs, allowing recruitment of other inflammatory cells into the CNS. T cells entering the CNS encounter their cognate myelin antigens and become reactivated by local antibody producing cells. T cells expand and release inflammatory mediators which help recruit other immune cells to the site of inflammation, in particular activated B cells. B cells and plasma cells accumulate in the perivascular and subarachnoid spaces and are thought to release demyelinating anti-myelin/oligodendrocyte antibodies into the surrounding tissue.[33] Activation of local microglial cells and infiltrating immune cells results in production of proteases, glutamate, reactive oxygen species and other cytotoxic agents which promote myelin breakdown. The same cytotoxic mediators are

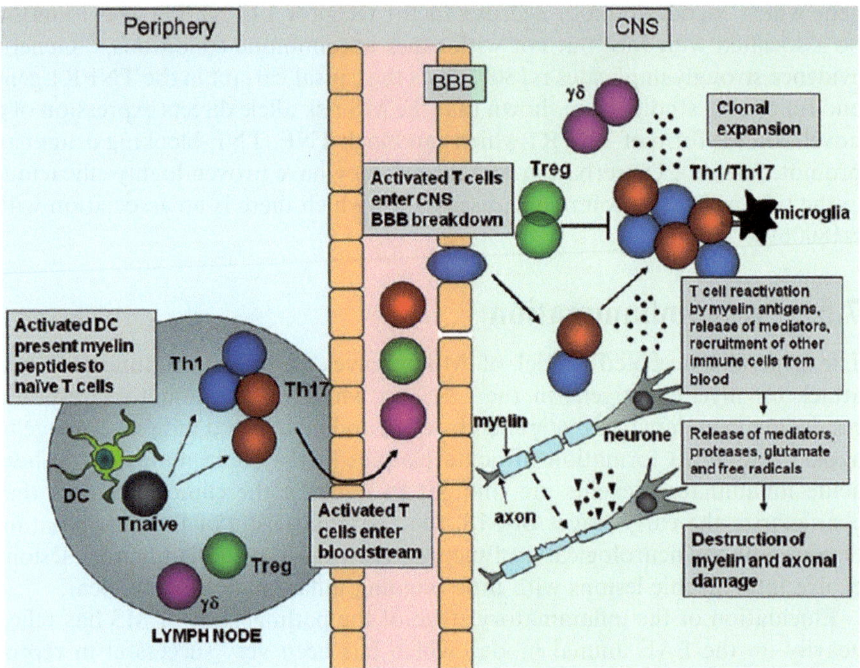

Figure 7.5   Migration and effector function of T cells in the CNS during EAE. After immunization with myelin antigens, complete Freund's adjuvant (CFA) and pertussis toxin, DCs are activated in the lymph nodes by TLR agonists within the mycobacterium tuberculosis component of CFA, and present myelin antigen to naive T cells. The activated myelin-specific T cells enter the bloodstream and traffic to and enter the CNS. Breakdown of the BBB occurs, allowing recruitment of other inflammatory cells into the CNS. T cells entering the CNS encounter their cognate myelin antigens and become reactivated by local APC. T cells expand and release inflammatory mediators which help recruit other immune cells to the site of inflammation. Activation of local microglial cells and infiltrating cells results in production of proteases, glutamate, reactive oxygen species and other cytotoxic agents which promote myelin breakdown. Damage to the myelin sheath surrounding axons is followed by axonal damage and neurological impairment.
(Reproduced from Fletcher *et al.*[31] with permission from Wiley).

thought to be responsible for the accompanying axon damage. The relative contribution of these various pathological events to the clinical deficit is unclear at early stages.

There is evidence of increasing axonal loss together with atrophy of the CNS as MS progresses, despite reduction in classic inflammatory markers, and this has led to the suggestion that there is an underlying neurodegenerative component that becomes the predominant pathology during the later stages.[34] More recent studies have provided evidence for extensive pathology in the grey matter, including demyelination, axon and neurite damage and neuronal loss, that often exceeds the extent of the pathology in the white matter.[35,36] The neuronal loss is largely independent of the demyelination,[37] indicating the presence of a neurodegenerative process. Both imaging and pathology studies suggest that diffuse grey matter pathology in the cerebral cortex may be a major driver of the progressive neurological deficit that characterizes the later stages of MS,[38,39] thus promoting the idea that a major neurodegenerative component exists together with the earlier inflammatory stage.

Clearly, activated microglia play an important role in the inflammatory processes in MS, since they are found in actively demyelinating lesions. Their role in the differentiation of T cells could lead to the expansion of inflammation and tissue destruction. However, microglia are also involved in the termination of an inflammatory response and produce protective factors. The most important microglia-associated molecules in MS, CD40, B7-1 and B7-2, interferon-γ, TNF, chemokines, prostanoids and nitric oxide, have been identified, but our ability to therapeutically manipulate them in microglia at the onset and in the development of MS needs further work. During the later progressive stages of MS, microglia continue to play an important role and it is suggested that they are largely responsible for the neuronal damage. However, as yet there is little experimental evidence available to substantiate this, as existing EAE models do not reproduce these later stages.

7.6　Metals Involved

Iron deposition in human brain tissue occurs in the process of normal aging and in many neurodegenerative diseases. Elevated iron levels in certain brain regions have also been noted in MS. These insights came originally from histological examinations and more recently from MRI. The exact mechanism(s) for this phenomenon and its implication in terms of pathophysiology and clinical significance are still largely unknown. In line with other neurodegenerative diseases, it has been suggested that iron-mediated oxidative stress could be responsible for the formation of cytotoxic protein aggregates, which might trigger or promote neurodegeneration, but such aggregates have not been demonstrated in the MS brain. The accumulation of iron could be due to the inflammatory processes involved in MS disrupting the BBB and attracting iron-rich macrophages, and inflammation might also be responsible for reducing axonal clearance of iron in MS brains. Existing data indicate that iron accumulation in MS is most pronounced in the deep grey matter structures and significantly extends beyond age-related effects. Iron accumulation in these regions seems to be associated with disease severity in terms of atrophy and disease duration.[40,41]

Excess levels of iron are represented concomitantly in multiple deep grey matter structures, often with bilateral representation, whereas in white matter, pathological iron deposits are usually located at sites of inflammation that are associated with veins. These distinct spatial patterns suggest disparate mechanisms of iron accumulation between these regions. Excess iron has been postulated to promote disease activity in MS by: (i) amplifying the activated state of microglia, resulting in increased production of proinflammatory mediators; (ii) promoting mitochondrial dysfunction; and (iii) catalysing the production of damaging reactive oxygen species (ROS).[42] The pathological consequences of abnormal iron deposits may be dependent on the affected brain region and/or the accumulation process. Because iron in the brain is predominantly stored in oligodendrocytes,[43] these cells and the myelin sheaths they produce may be more susceptible to an iron-mediated increase in ROS. In addition, the death of oligodendrocytes associated with the demyelination would also release Fe^{2+} into the extracellular milieu that might then amplify oxidative stress in axons.[44]

Trace metal analysis in cerebrospinal fluid (CSF) from MS patients indicated that copper levels were significantly elevated, manganese levels were significantly decreased and there were no significant differences in zinc levels compared to controls.[45]

7.7 Therapeutics

On the basis of a number of criteria, the treatment therapies which at present are used for MS are not satisfactory. In this concluding section, we begin by briefly reviewing current approaches before passing to review some of the preclinical approaches and therapies in clinical trials which have been developed as a result of improved understanding of the pathophysiology of MS. Therapeutic approaches to MS have been recently reviewed.[46,47] A schematic representation of MS pathophysiology, indicating the sites at which treatment intervention may act, is presented in Figure 7.6.

The original treatments were based on type I interferons (IFs), without any real idea concerning how they might be interfering with the disease process. Interferon-β reduces the rate of MS relapses by a third compared to placebo. With time, we have learnt that interferons can divert the cytokine milieu towards a less inflammatory phenotype, reflected in increased levels of anti-inflammatory IL-4 and IL-10, coincident with a decrease in inflammatory mediators such as IL-17, IL-23 and osteopontin. It is now generally accepted that IFs divert the immune system away from a proinflammatory Th1 and Th17 response towards a more anti-inflammatory Th2 one. IFs may also act by expanding regulatory cells like NK cells and T-regulatory cells (T-regs). In addition to effects on immune cells, IFs have also been shown to affect a number of adhesion molecules and matrix metalloproteinases, which can result in impaired migration and trafficking of lymphocytes into the CNS. IFs have also been found to affect B cell function, modulating cytokine secretion.[48]

Glatiramer acetate (GA), like IFs, has been used for some time in the therapy of MS, since its approval for relapsing–remitting MS (RRMS) in the USA in 1996. GA is a polymer made up of a random sequence of the four most frequent amino acids found in myelin basic protein (MBP) originally designed to induce EAE; it in fact turned out to suppress the disease in the animal model.[49] Like IF, GA reduces the rate of MS relapses by about a third, acting on cells of both the innate and adaptive immune system, modulating both T cell and B cell

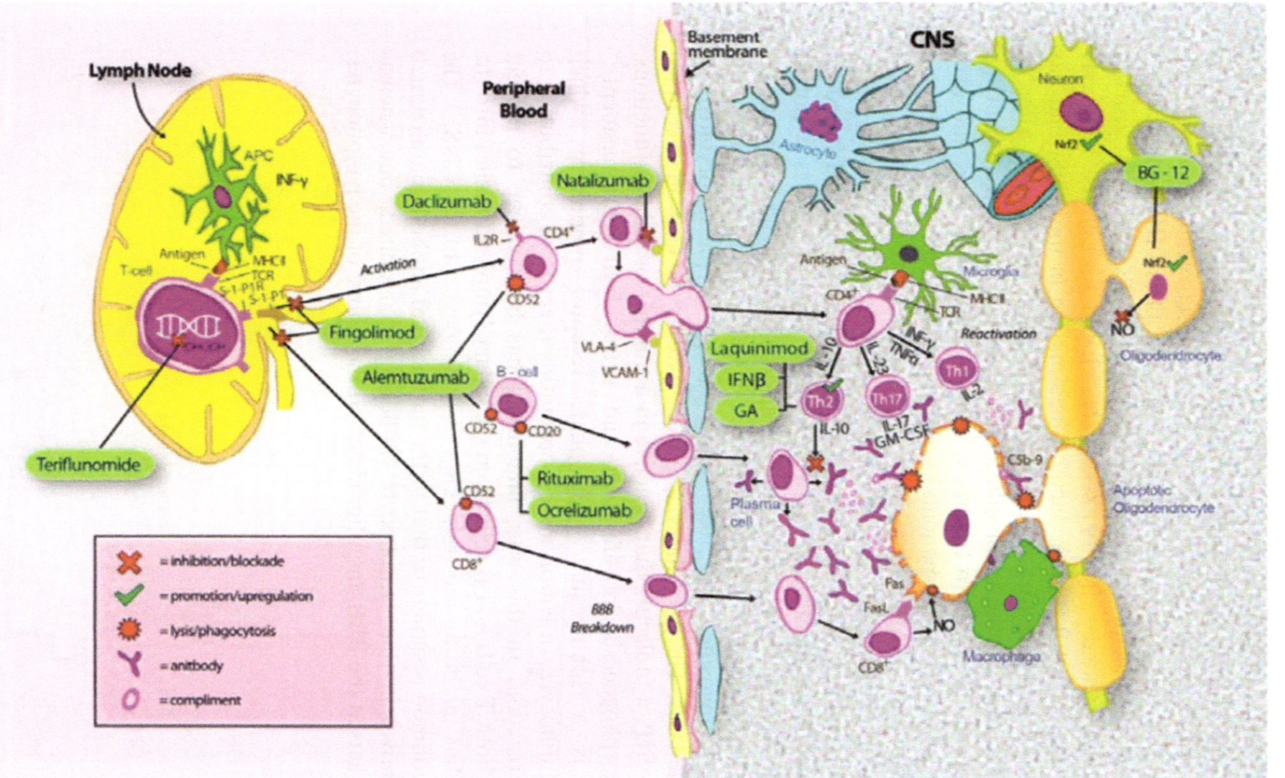

Figure 7.6 Schematic representation of MS pathophysiology, indicating points of treatment intervention. APC, antigen presenting cell; BBB, blood–brain barrier; C5b-9, complement complex 5b-9; CNS, central nervous system; DHODH, dihydroorotate dehydrogenase; FasL, Fas ligand; GA, glatiramer acetate; IFN, interferon; IL2R, interleukin 2 receptor; MHC I, major histocompatibility complex I; NO, nitrous oxide; Nrf2, nuclear factor (erythrocyte derived) related factor 2; S-1-P1, sphingosine-1-phosphate 1; TCR, T cell receptor; VCAM-1, vascular cell adhesion molecule 1; VLA-4, very late antigen 4. (Reproduced from Buzzard *et al.*[46] with permission from MDPI).

function. Recent results suggest that type II (anti-inflammatory) monocytes may be one of the principal targets of GA, promoting differentiation of naïve $CD4^+$ T cells into Th2 cells and T-reg cells, with a reciprocal decrease in production of Th1 and Th17 subsets.[50]

Natalizumab (NTZ; Tysabri®) is a humanized monoclonal antibody, and is the first alpha4 integrin antagonist in a new class of selective adhesion-molecule inhibitors. It was first approved in 2007 following phase III studies which revealed robust evidence of a reduction in relapses and disability at 2 years in RRMS patients treated with NTZ.[51,52] Natalizumab inhibits leukocyte migration across the BBB, thus reducing inflammation in the CNS, and has been approved worldwide for the treatment of RRMS. Five years after its introduction, it remains the most efficacious of all approved treatments in RRMS. However, natalizumab treatment is not without some risk, and over 200 cases (0.25% of treated patients worldwide) of progressive multifocal leukoencephalopathy (PML) have been reported in patients treated with natalizumab,[53] of which approximately 1 in 5 were fatal.[54]

Fingolimod (Gilenya, FTY720; Novartis), a synthetic compound based on the fungal secondary metabolite myriocin (ISP-I) (Figure 7.7), is a potent immunosuppressant that was approved in 2010–2011 as the first oral treatment for multiple sclerosis (MS).[55] Fingolimod has a similar chemical structure to the sphingolipid sphingosine, and like sphingosine it is phosphorylated by sphingosine kinases. The phosphorylated form inhibits the G-protein-coupled receptors for S1P, a molecule that is active in a number of cellular functions, including morphogenesis, proliferation and cytoskeletal rearrangement.[56] Fingolimod is a sphingosine 1-phosphate receptor modulator, which sequesters lymphocytes in lymph nodes, preventing them from contributing to an immune reaction. In this way, fingolimod has been shown to prevent the egress of both T cells and B cells from secondary lymphoid tissues into the circulation, thus inhibiting the trafficking of lymphocytes to the CNS.[57] The action of fingolimod is similar to that of the monoclonal antibody natalizumab (discussed above) in that lymphocyte trafficking is inhibited, but whereas natalizumab acts at the BBB, fingolimod acts at the source of the lymphocytes.

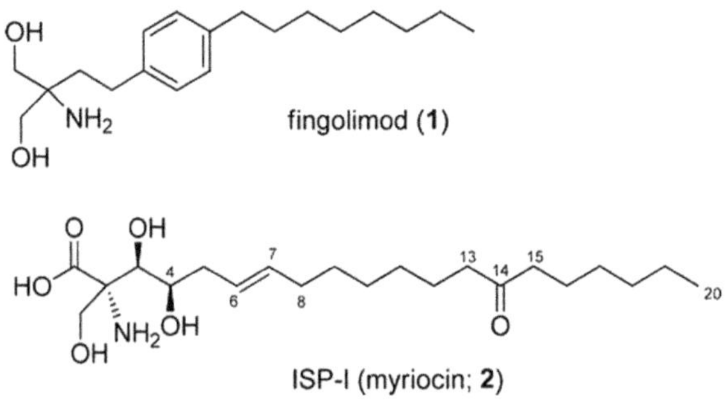

fingolimod (**1**)

ISP-I (myriocin; **2**)

Figure 7.7 Structures of fingolimod (**1**) and of myriocin (ISP-I) (**2**).

Dihydroorotate dehydrogenase (DHODH) is a mitochondrial enzyme that catalyses the fourth step in the pathway of pyrimidine biosynthesis, namely the conversion of dihydroorotate to orotate. Inhibitors of DHODH exhibit anti-inflammatory and immunosuppressant properties and have been shown to be partially effective in the treatment of autoimmune diseases such as MS.[58] Leflunomide (Figure 7.8) is a low molecular weight inhibitor of dihydro-orotate dehydrogenase and is a treatment with recognized efficacy and relative safety in the management of rheumatoid arthritis (RA). Teriflunomide (Sanofi-Aventis) is the active metabolite of leflunomide and is an oral dihydroorotate dehydrogenase inhibitor that inhibits T cell proliferation.[59] It was approved by the FDA in 2012 for the treatment of MS. By blocking pyrimidine synthesis, cell division is inhibited in rapidly dividing cells such as T cells. Furthermore, in animal studies, teriflunomide was shown to suppress tyrosine kinases involved in signal transduction pathways, thus altering calcium mobilization and T cell activation.[60]

Recently, there has been a shift in MS therapies from parenteral treatment to oral therapy, in an attempt to improve patient acceptance and compliance. Esters of fumaric acid, an intermediate of the citric acid cycle, such as dimethyl fumarate and monomethyl fumarate (fumaderm), administered orally, have been used in the treatment of psoriasis for decades in German-speaking countries,[61] although the mechanism of action is not completely understood. It was shown that dimethyl fumarate was the principal component that provided antipsoriatic activity, and this led to the development of BG-12 (Biogen-Idec), dimethyl fumarate, a second-generation formulation of fumaderm,[62] that is currently awaiting FDA approval. Fumaric acid esters have several mechanisms of action relevant to the treatment of MS, notably by reducing oxidative stress *via* a reduction in the nuclear translocation of nuclear factor kappa-light-chain enhancer of activated B cells (NF-κB). NF-κB plays an important role as a transcription factor for a variety of proinflammatory mediators such as cytokines, chemokines and adhesion factors.[63] A reduction in NF-κB activity leads to the decreased expression of an array of proin-flammatory molecules, which in turn can have dramatic effects on subsets of inflammatory cells. They can induce a shift in the cytokine profile to favour the anti-inflammatory Th2 cytokines rather than the proinflammatory Th1 pattern.[64] Th2 cytokines induce apoptosis in activated T cells and reduce the expression of adhesion molecules.[65–67] In the EAE animal model, a fumaric acid ester has been shown to reduce the severity of disease by increasing the expression of the anti-inflammatory cytokine IL-10 and decreasing the expression of proinflammatory cytokines TNFα and IL-6.[68]

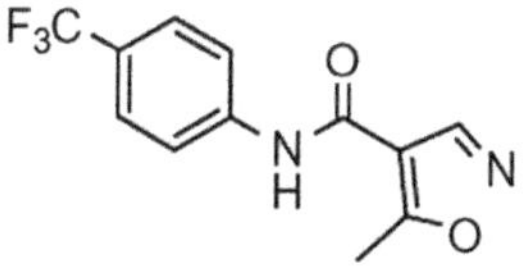

Figure 7.8   Structure of teriflunomide.

Laquinimod (Teva Pharmaceuticals and Active Biotech) is an oral immunomodulator which is structurally similar to roquinimex (Linomide) (Figure 7.9). Roquinimex was shown to reduce the degree of new lesions in phase 2 and phase 3 MS studies, but its development was discontinued because of adverse effects, including serositis and myocardial infarction.[69–72] More than 50 derivatives of roquinimex were synthesized in an effort to reduce its toxicity and improve its efficacy and then tested in the mouse EAE model for efficacy. Selected compounds were further tested for toxicity in the beagle for induction of a proinflammatory reaction;[73] from this screen, laquinimod was chosen. In animal models, laquinimod has been shown to inhibit T cell and macrophage entry into the CNS and to shift the cytokine profile of T cells from Th1 to Th2/Th3.[74] In phase III clinical trials,[75] laquinimod was unable to significantly reduce relapses into MS among patients beyond a placebo. However, taken together with the improvement in disability progression, it is possible that laquinimod may be beneficial in reducing the neurodegenerative component of MS.

Cladribine (Figure 7.10) is a purine analogue which is phosphorylated by deoxycytidine kinase, but, because it is resistant to deamination, it achieves high intracellular concentrations[76] and it is thought that its accumulation interferes with DNA repair, thus damaging DNA and resulting in cell death. Lymphocytes, in particular, have high concentrations of deoxycytidine kinase and, therefore, are relatively selectively sensitive to cell death by cladribine.[77]

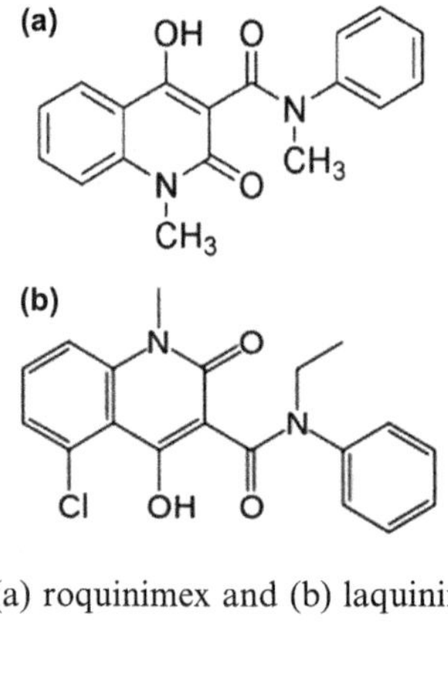

Figure 7.9   Structures of (a) roquinimex and (b) laquinimod.

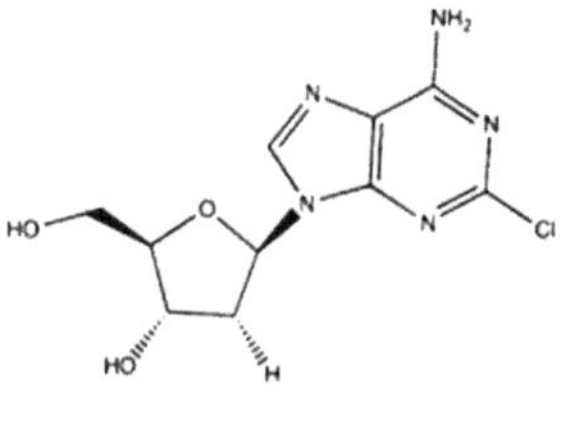

Figure 7.10   Structure of cladribine.

Because of this lymphocyte selectivity, it was theorized that cladribine might be effective in T and B cell-mediated autoimmune disorders. The first indication that this theory may be correct came from results in which about 50% of patients with treatment-resistant autoimmune hemolytic anaemia responded to cladribine.[76] Clinical trials have, to date, been disappointing. There are likely to be long-term adverse effects that would increase neurodegeneration rather than improve it, such as reduced DNA repair and reduced remyelination due to inhibition of cell division.

Alemtuzumab is a humanized monoclonal antibody that targets CD52, a glycoprotein cell surface marker of unknown function found on all differentiated lymphocytes and monocytes but not on hematopoietic precursors.[78] Treatment with alemtuzumab results in the rapid depletion of CD52-expressing cells *via* antibody-dependent cellular cytotoxicity. However, up to 30% of alemtuzumab-treated patients developed immune-mediated diseases, with significantly higher levels of IL-21 and lower levels of IL-7.[79]

Daclizumab is a humanized monoclonal antibody directed against the alpha chain of the IL-2 receptor (CD25). Daclizumab binds to the IL-2 binding site of CD25, masking the binding site without influencing signalling activities and without causing cytotoxicity.[80] The interaction of IL-2 with its receptor expands the generation of effector and regulatory T cells,[81] and also expands the population of NK cells. This expansion of NK cells is thought to be due to the increased bioavailability of IL-2 to NK cells, which in turn stimulates NK cell proliferation through binding to the IL-2 intermediate affinity receptor.[82]

Rituximab is a chimeric mouse–human monoclonal antibody designed to target B cells expressing the CD20 transmembrane protein. Stem cells, pro-B cells and differentiated plasma cells are exempt from the effects of rituximab owing to the fact that they do not express CD20. Rituximab efficiently removes $CD20^+$ B cells from the circulation and as such it has proven beneficial in the treatment of B cell lymphoma. Rituximab lyses B cells *via* complement-dependent cytotoxicity, antibody-dependent cytotoxicity and B cell apoptosis.[83] Treatment with rituximab in MS patients is well tolerated and risks appear low over relatively short exposures, and treatment efficiently depletes circulating B cells and results in substantial decreases in new MS disease activity.[83]

Ocrelizumab is a recombinant humanized monoclonal antibody that selectively targets $CD20^+$ B cells. While sharing a significant degree of sequence homology with rituximab, it binds to a distinct region of CD20,[84] and appears to be associated with greater antibody-dependent cytotoxicity and less complement-mediated cytotoxicity than rituximab.[85]

Atacicept is a novel therapy which endeavours to exploit previously reported beneficial effects of B cell targeted therapy in MS. Atacicept is a human recombinant fusion protein comprising the binding region of a receptor that binds two cytokines, a proliferation-inducing ligand (APRIL) and B lymphocyte stimulator (BLyS), members of the TNF superfamily, which are expressed by cells of both the innate and adaptive immune systems. These cytokines are involved in the regulation of B cell development, homeostasis and survival,[86] and both are increased in a number of autoimmune-mediated diseases, including MS.[87–90] Atacicept acts as an antagonist, blocking the binding of both APRIL and BLyS to their receptors and inhibiting the downstream effects on B cells. Owing to the variable expression of the receptors in B cell subsets, atacicept affects mature B cells and antibody producing plasma cells but spares B cell progenitors and memory B cells.[87,91] Despite achieving the expected reduction in circulating B cells and immunoglobulin levels in atacicept-treated animals,[92] atacicept has failed as a treatment for RRMS on account of the increased CNS inflammatory activity seen in atacicept-treated RRMS patients.

In EAE, TNF is clearly proinflammatory during the acute phases of the disease and demyelination is more extensive when TNF is present, and a large body of evidence from studies in MS and EAE initially suggested that neutralization of TNF activity could benefit the course of the disease. A number of different anti-TNF antibodies are available on the market to treat other autoimmune diseases,[25] but the administration of such anti-TNF agents to MS patients resulted in the opposite effect, increasing clinical and radiological disease activity in the patients. However, both clinical trials involved agents that blocked the action of both soluble and transmembrane TNF and it has been suggested[25] that agents that selectively inhibit sTNF or signals from TNFR1 without blocking the binding of tmTNF to TNFR2 could be effective in the progressive phase of MS.

In preclinical MS models, inhibitors of poly(ADP-ribose) polymerase (PARP)-1 have shown great promise. Emerging evidence demonstrates that PARP-1 inhibitors epigenetically regulate gene expression and finely tune transcriptional activation in immune and neural cells. It has been proposed that drugs inhibiting PARP-1 activity protect from neuroinflammation in MS models *via* immunomodulation and direct neuroprotection.[93]

Despite decades of research, current treatments for RRMS are unable to induce complete remission in all patients, and have little impact on disease progression. The response of individual patients to a specific treatment is highly variable, which probably reflects heterogeneity in terms of the underlying immunopathogenesis. Clearly, migration of peripheral lymphocytes and monocytes into the CNS is a critical step in the instigation and propagation of the demyelination and tissue damage found in MS. Indeed, one relative success in MS treatment strategies has been drugs which target leucocyte migration. Both natulizumab and fingolimod were designed to impede trafficking of lymphocytes to the CNS, and it has been shown that other treatments such as IFNβ, rituximab, teriflunomide and laquinimod exert effects on lymphocyte migration, usually by modulation of chemokine and adhesion molecule expression.[94–98] Other targets include increasing levels of neurotrophic factors like BDNF (which has been shown for laquinimod, IFBβ and GA),[99–105] reducing oxidative stress (BG-12) and exerting direct effects on astrocytes (fingolimod).[106] As our understanding of the complex pathogenesis of MS grows, we can hope that additional targeted treatments will be developed.

References

1. H. Lassmann, W. Brück and C. Lucchinetti, *Trends Mol. Med.*, 2001, **7**, 115.
2. A. Compston and A. Coles, *Lancet*, 2002, **359**, 1221.
3. A. Compston and A. Coles, *Lancet*, 2008, **372**, 1502.
4. W. I. McDonald, A. Compston, G. Edan, D. Goodkin, H. P. Hartung, F. D. Lublin, H. F. McFarland, D. W. Paty, C. H. Polman, S. C. Reingold, M. Sandberg-Wollheim, W. Sibley, A. Thompson, S. van den Noort, B. Y. Weinshenker and J. S. Wolinsky, *Ann. Neurol.*, 2001, **50**, 121.
5. S.E. Baranzini, *Curr. Opin. Genet. Dev.*, 2011, **21**, 317.
6. The International Multiple Sclerosis Genetics Consortium (IMSGC), *Am. J. Hum. Genet.*, 2010, **86**, 621.
7. R. Hohlfeld and H. Wekerle, *Proc. Natl. Acad. Sci. U. S. A.*, 2004, **101**, 14599.
8. H. Lassmann, *Comparative Neuropathology of Chronic Experimental Allergic Encephalomyelitis and Multiple Sclerosis*, Springer, Berlin, 1983, pp. 1–135.
9. R. Gold, C. Linington and H. Lassmann, *Brain*, 2006, **129**, 1953.
10. A. Ascherio, K. L. Munger and J. D. Lünemann, *Nat. Rev. Neurol.*, 2012, **8**, 602.
11. D. A. Dyment, I. M. Yee, G. C. Ebers and A. D. Sadovnik, *J. Neurol. Neurosurg. Psychiatry*, 2006, **77**, 258.
12. K. Hemminski, X. Li, J. Sundquist, J. Hillert and K. Sundquist, *Neurogenetics*, 2009, **10**, 5.
13. International Multiple Sclerosis Genetics Consortium (IMSGC), *Am. J. Hum. Genet.*, 2005, **77**, 454.
14. P. A. Gourraud, H. F. Harbo, S. L. Hauser and S. E. Baranzini, *Immunol. Rev.*, 2012, **248**, 87.
15. A. Kemppinen, S. Sawcer and A. Compson, *Brief Funct. Genomics*, 2011, **10**, 61.
16. International Multiple Sclerosis Genetics Consortium (IMSGC) and Wellcome Trust Case Control Consortium (WTCCC 2), *Nature*, 2011, **476**, 214.
17. C. Du and X. Xie, *Cell Res.*, 2012, **22**, 1108.
18. J. Chun and V. Brinkmann, *Discov. Med.*, 2011, **12**, 213.
19. A. Chan, R. Gold and N. von Ahsen, *Curr. Pharm. Des.*, 2011, **17**, 2803.
20. M. Gambuzza, N. Licata, E. Palella, D. Celi, V. Foti Cuzzola, D. Italiano, S. Marino and P. Bramanti, *J. Neuroimmunol.*, 2011, **239**, 1.
21. T. Okuno, Y. Nakatsuji and A. Kumanogoh, *FEBS Lett.*, 2011, **585**, 3829.
22. M. Braitch and C. S. Constantinescu, *Inflammation Allergy: Drug Targets*, 2010, **9**, 249.
23. K. Morimoto, S. Kon, Y. Matsui and T. Uede, *Curr. Drug Targets*, 2010, **11**, 494.

24. L. De Santi, G. Polimeni, S. Cuzzocrea, E. Esposito, E. Sessa, P. Annunziata and P. Bramanti, *Curr. Med. Chem.*, 2011, **18**, 1775.
25. A. Caminero, M. Comabella and X. Montalban, *J. Neuroimmunol.*, 2011, **234**, 1.
26. M. K. McCoy and M. G. Tansey, *J. Neuroinflammation*, 2008, **5**, 45.
27. B. Beutler and A. Cerami, *Annu. Rev. Immunol.*, 1989, **7**, 625.
28. A. P. Gregory, C. A. Dendrou, K. E. Attfield, A. Haghikia, D. K. Xifara, F. Butter, G. Poschmann, G. Kaur, L. Lambert, O. A. Leach, S. Prömel, D. Punwani, J. H. Felce, S. J. Davis, R. Gold, F. C. Nielsen, R. M. Siegel, M. Mann, J. L. Bell, G. McVean and F. Fugger, *Nature*, 2012, **488**, 508.
29. E. M. Frohman, M. K. Racke and C. S Raine, *New Eng. J. Med.*, 2006, **354**, 942.
30. International Multiple Sclerosis Genetics Consortium, *New Eng. J. Med.*, 2007, **357**, 851.
31. J. M. Fletcher, S. J. Lalor, C. M. Sweeney, N. Tubridy and K. H. Mills, *Clin. Exp. Immunol.*, 2010, **162**, 1.
32. F. Petermann and T. Korn, *FEBS Lett.*, 2011, **585**, 3747.
33. A. Uccelli, F. Aloisi and V. Pistoia, *Trends Immunol.*, 2005, **26**, 254.
34. L Steinman, *Nat. Immunol.*, 2001, **2**, 762.
35. A. Kutzelnigg, C. F. Lucchinetti, C. Stadelmann, W. Bruck, H. Rauschka, M. Bergmann, M. Schmidbauer, J. E. Parisi and H. Lassmann, *Brain*, 2005, **128**, 2705.
36. R. Magliozzi, O. Howell, A. Vora, B. Serafini, R. Nicholas, M. Puopolo, R. Reynolds and F. Aloisi, *Brain*, 2007, **130**, 1089.
37. R. Magliozzi, O. W. Howell, C. Reeves, F. Roncaroli, R. Nicholas, B. Serafini, F. Aloisi and R. Reynolds, *Ann. Neurol.*, 2010, **68**, 477.
38. M. Calabrese, M. Filippi and P. Gallo, *Nat. Rev. Neurol.*, 2010, **6**, 438.
39. R. Reynolds, F. Roncaroli, R. Nicholas, B. Radotra, D. Gveric and O. Howell, *Acta Neuropathol.*, 2011, **122**, 155.
40. M. Neema, A. Arora, B. C. Healy, Z. D. Guss, S. D. Brass, Y. Duan, G. J Buckle, B. I. Glanz, L. Stazzone, S. J. Khoury, H. L. Weiner, C. R. Guttmann and R. Bakshi, *J. Neuroimaging*, 2009, **19**, 3.
41. S. Ropele, W. de Graaf, M. Khalil, M. P. Wattjes, C. Langkammer, M. A. Rocca, A. Rovira, J. Palace, F. Barkhof, M. Filippi and F. Fazekas, *J. Magn. Reson. Imaging*, 2011, **34**, 13.
42. R. Williams, C. L. Buchheit, N. E. Berman and S. M. LeVine, *J. Neurochem.*, 2012, **120**, 7.
43. S. W. Hulet, S. Powers and J. R. Connor, *J. Neurol. Sci.*, 1999, **165**, 48.
44. H. Lassmann, J. Van Horssen and D. Mahad, *Nat. Rev. Neurol.*, 2012, **8**, 647.
45. T. M. Melø, C. Larsen, L. R. White, J. Aasly, T. E. Sjøbakk, T. P. Flaten, U. Sonnewald and T. Syversen, *Biol. Trace Elem. Res.*, 2003, **93**, 1.
46. K. A. Buzzard, S. A. Broadley and H. Butzkeuven, *Int. J. Mol. Sci.*, 2012, **13**, 12665.
47. P. Fontoura and H. Garren, *Results Probl. Cell Differ.*, 2010, **51**, 259.

48. V. S. Ramgolam, Y. Sha, K. L. Markus, N. Choudhary, L. Troiani, M. Chopra and B. Marcovic-Plesa, *J. Immunol.*, 2011, **186**, 4518.

49. M. Sela and D. Teitelbaum, *Expert Opin. Pharmacother.*, 2001, **2**, 1149.

50. P. H. Lalive, O. Neuhaus, M. Benkhoucha, D. Burger, R. Hohlfeld, S. S. Zamvil and M. S. Weber, *CNS Drugs*, 2011, **25**, 401.

51. C. H. Polman, P. W. O'Connor, E. Havrdova, M. Hutchinson, L. Kappos, D. H. Miller, J. T. Phillips, F. D. Lublin, G. Giovannoni, A. Wajgt, M. Toal, F. Lynn, M. A. Panzara, A. W. Sandrock and AFFIRM Investigators, *New Engl. J. Med.*, 2006, **354**, 899.

52. E. Pucci, G. Giuliani, A. Solari, S. Simi, S. Minozzi, C. Di Pietrantonj and I. Galea, *Cochrane Database Syst. Rev.*, 2011, CD007621.

53. G. Bloomgren, S. Richman, C. Hotermans, M. Subramanyam, S. Goelz, A. Natarajan, S. Lee, T. Plavina, J. V. Scanlon, A. Sandrock and C. Bozic, *New Engl. J. Med.*, 2012, **366**, 1870.

54. Biogen Idec Tysabri Safety Update, available online (accessed on 17 August, 2012): http://www.tapp.com.au/members/Tysabri_Safety_Update_160812.pdf.

55. C. R. Strader, C. J. Pearce and N. H. Oberlies, *J. Nat. Prod.*, 2011, **74**, 900.

56. V. Brinkmann, *Pharmacol. Ther.*, 2007, **115**, 84.

57. L. Kappos, J. Antel, G. Comi, X. Montalban, P. O'Connor, C. H. Polman, T. Haas, A. A. Korn, G. Karlsson and E. W. Radue, *New Engl. J. Med.*, 2006, **355**, 1124.

58. P. W. O'Connor, D. Li, M. S. Freedman, A. Bar-Or, G. P. A. Rice, C. Confraveux, D. W. Patty, J. A. Steweart and R. Scheyer, *Neurology*, 2006, **66**, 894.

59. H. M. Cherwinski, D. McCarley, R. Schatzman, B. Devens and J. T. Ransom, *J. Pharmacol. Exp. Ther.*, 1995, **272**, 460.

60. T. Korn, K. Toyka, H. P. Harting and S. Jung, *Brain*, 2001, **124**, 1791.

61. P. J. Altmeyer, U. Matthes, F. Pawlak, K. Hoffmann, P. J. Frosch, P. Ruppert, S. W. Wassilew, T. Horn, H. W. Kreysel and G. Lutz, *et al*, *J. Am. Acad. Dermatol.*, 1994, **30**, 977.

62. M. Wakkee and H. B. Thio, *Curr. Opin. Invest. Drugs*, 2007, **8**, 955.

63. M. S. Hayden and S. Ghosh, *Genes Dev.*, 2012, **26**, 203.

64. R. De Jong, A. C. Bezemer, T. P. Zomerdijk, T. van de Pouw-Kraan, T. H. Ottenhoff and P. H. Nibbering, *Eur. J. Immunol.*, 1996, **26**, 2067.

65. U. Mrowietz and K. Asadullah, *Trends Mol. Med.*, 2005, **11**, 43.

66. H. M. Ockenfels, T. Schultewolter, G. Ockenfels, R. Funk and M. Goos, *Br. J. Dermatol.*, 1998, **139**, 390.

67. M. Vandermeeren, S. Janssens, M. Borgers and J. Geysen, *Biochem. Biophys. Res. Commun.*, 1997, **234**, 19.

68. S. Schilling, S. Goelz, R. Linker, F. Luehder and R. Gold, *Clin. Exp. Immunol.*, 2006, **145**, 101.

69. O. Andersen, J. Lycke, P. O. Tollesson, A. Svenningsson, B. Runmarker, A. S. Linde, M. Aström, P. Gjörstrup and S. Ekholm, *Neurology*, 1996, **47**, 895.

70. D. M. Karussis, Z. Meiner, D. Lehmann, J. M. Gomori, A. Schwarz, A. Linde and O. Abramsky, *Neurology*, 1996, **47**, 341.

71. J. H. Noseworthy, J. S. Wolinsky, F. D. Lublin, J. H. Whitaker, A. Linde, P. Gjorstrup and H. S. Sullivan, *Neurology*, 2000, **54**, 1726.

72. I. L. Tan, G. J. Lycklama à Nijeholt, C. H. Polman, H. J. Adèr and F. Barkhof, *Mult. Scler.*, 2000, **6**, 99.

73. S. Jönsson, G. Andersson, T. Fex, T. Fristedt, G. Hedlund, K. Jansson, L. Abramo, I. Fritzson, O. Pekarski, A. Runström, H. Sandin, I. Thuvesson and A. Björk, *J. Med. Chem.*, 2004, **47**, 2075.

74. J. S. Yang, L. Y. Xu, B. G. Xiao, G. Hedlund and H. Link, *J. Neuroimmunol.*, 2004, **156**, 3.

75. G. Comi, D. Jeffery, L. Kappos, X. Montalban, A. Boyko, M. A. Rocca and M. Filippi, *New Engl. J. Med.*, 2012, **366**, 1000.

76. E. Beutler, *Lancet*, 1992, **340**, 952.

77. S. Seto, C. J. Carrera, M. Kubota, D. B. Wasson and D. A. Carson, *J. Clin. Invest.*, 1985, **75**, 377.

78. M. H. Gilleece and T. M. Dexter, *Blood*, 1993, **82**, 807.

79. J. L. Jones, C. L. Phuah, A. L. Cox, S. A. Thompson, M. Ban, J. Shawcross, A. Walton, S. J. Sawcer, A. Compston and A. J. Coles, *J. Clin. Invest.*, 2009, **119**, 2052.

80. A. Lutterotti and R. Martin, *Lancet Neurol.*, 2008, **7**, 538.

81. H. P. Kim, J. Imbert and W. J. Leonard, *Cytokine Growth Factor Rev.*, 2006, **17**, 349.

82. J. F. Martin, J. S. Perry, N. R. Jakhete, X. Wang and B. Bielekova, *J. Immunol.*, 2010, **185**, 1311.

83. B. Barun and A. Bar-Or, *Clin. Immunol.*, 2012, **142**, 31.

84. M. C. Genovese, J. L. Kaine, M. B. Lowenstein, J. Giudice, A. Baldassare, J. Schechtman, E. Fudman, M. Kohen, S. Gujrathi and R. G. Trapp, *et al*, *Arthritis Rheum.*, 2008, **58**, 2652.

85. L. Kappos, D. Li, P. A. Calabresi, P. O'Connor, A. Bar-Or, F. Barkhof, M. Yin, D. Leppert, R. Glanzman, J. Tinbergen and S. L. Hauser, *Lancet*, 2011, **378**, 486.

86. H. P. Hartung and B. C. Kieseier, *Ther. Adv. Neurol. Disord.*, 2010, **3**, 205.

87. S. R. Dillon, J. A. Gross, S. M. Ansell and A. J. Novak, *Nat. Rev. Drug Discov.*, 2006, **5**, 235.

88. M. Thangarajh, A. Gomes, T. Masterman, J. Hillert and P. Hjelmstrom, *J. Neuroimmunol.*, 2004, **152**, 183.

89. M. Thangarajh, T. Masterman, J. Hillert, S. Moerk and R. Jonsson, *Scand. J. Immunol.*, 2007, **65**, 92.

90. M. Thangarajh, T. Masterman, U. Rot, K. Duvefelt, B. Brynedal, V. D. Karrenbauer and J. Hillert, *J. Neuroimmunol.*, 2005, **167**, 210.

91. J. A. Gross, S. R. Dillon, S. Mudri, J. Johnston, A. Littau, R. Roque, M. Rixon, O. Schou, K. P. Foley, H. Haugen, *et al*, *Immunity*, 2001, **15**, 289.

92. M. Carbonatto, P. Yu, M. Bertolino, E. Vigna, S. Steidler, L. Fava, C. Daghero, B. Roattino, M. Onidi, M. Ardizzone, *et al*, *Toxicol. Sci.*, 2008, **105**, 200.

93. L. Cavone and A. Chiarugi, *Trends Mol. Med.*, 2012, **18**, 92.

94. M. Caggiula, A. P. Batocchi, G. Frisullo, F. Angelucci, A. K. Patanella, C. Sancricca, V. Nociti, P. A. Tonali and M. Mirabella, *Clin. Immunol.*, 2006, **118**, 77.

95. M. Zeyda, M. Poglitsch, R. Geyeregger, J. S. Smolen, G. J. Zlabinger, W. H. Horl, W. Waldhausl, T. M. Stulnig and M. D. Saemann, *Arthritis Rheum.*, 2005, **52**, 2730.

96. T. Korn, T. Magnus, K. Toyka and S. J. Jung, *Leukoc. Biol.*, 2004, **76**, 950.

97. C. Wegner, C. Stadelmann, R. Pfortner, E. Raymond, S. Feigelson, R. Alon, B. Timan, L. Hayardeny and W. J. Bruck, *Neuroimmunology*, 2010, **227**, 133.

98. L. Piccio, R. T. Naismith, K. Trinkaus, R. S. Klein, B. J. Parks, J. A. Lyons and A. H. Cross, *Arch. Neurol.*, 2010, **67**, 707.

99. M. Caggiula, A. P. Batocchi, G. Frisullo, F. Angelucci, A. K. Patanella, C. Sancricca, V. Nociti, P. A. Tonali and M. Mirabella, *Clin. Immunol.*, 2006, **118**, 77.

100. K. Biernacki, J. P. Antel, M. Blain, S. Narayanan, D. L. Arnold and A. Prat, *Arch. Neurol.*, 2005, **62**, 563.

101. P. Sarchielli, M. Zaffaroni, A. Floridi, L. Greco, A. Candeliere, A. Mattioni, S. Tenaglia, M. di Filippo and P. Calabresi, *Mult. Scler.*, 2007, **13**, 313.

102. Y. Blanco, E. A. Moral, M. Costa, M. Gomez-Choco, J. F. Torres-Peraza, L. Alonso-Magdalena, J. Alberch, D. Jaraquemada, T. Arbizu, F. Graus, *et al*, *Neurosci. Lett.*, 2006, **406**, 270.

103. R. Aharoni, R. Eilam, H. Domev, G. Labunskay, M. Sela and R. Arnon, *Proc. Natl. Acad. Sci. U. S. A.*, 2005, **102**, 19045.

104. D. Azoulay, V. Vachapova, B. Shihman and A. Miler, *J. Neuroimmunol.*, 2005, **167**, 215.

105. J. Thone, G. Ellrichmann, S. Seubert, I. Peruga, D. H. Lee, R. Conrad, L. Hayardeny, G. Comi, S. Wiese and R.A. Linker, *et al*, *Am. J. Pathol.*, 2012, **180**, 267.

106. J. W. Choi, S. E. Gardell, D. R. Herr, R. Rivera, C. W. Lee, K. Noguchi, S. T. Teo, Y. C. Yung, M. Lu, G. Kennedy, *et al*, *Proc. Natl. Acad. Sci. U. S. A.*, 2011, **108**, 751.

107. D. Voet and J. G. Voet, *Biochemistry*, Wiley, Hoboken, NJ, 4th edn, 2011.

Alcoholic Brain Damage

ROBERTA J. WARD

Catholique University of Louvain, Belgium
Email: roberta.ward@uclouvain.be

8.1 Introduction

Chronic alcoholic brain damage is induced by excessive consumption of alcohol. Even moderate alcohol consumers, with no specific neurological or hepatic problems, may show some evidence of regional brain damage and cognitive dysfunction. Multiple epidemiological reports have demonstrated that there is a U-shaped relationship between ethanol intake and general mortality,[1] the lowest death rate correlating with low to moderate amounts of ethanol, 1–3 drinks per day (15–45 g),[2] while abstaining from ethanol or excessive drinking was associated with higher mortality.[3] Moderate ethanol consumption may be beneficial to health: lower cardiovascular diseases[1] diminish the risk of dementia and Alzheimer's disease in the elderly[4] and may be protective against some forms of cancer.[5] Such beneficial effects of ethanol were initially ascribed to the presence of polyphenols, which are rich in alcoholic beverages (in red wine, quercertin, rutin, catechin, epicatechin and resveratrol) and minerals in beer (trace metals and vitamins), although only trace amounts are present. The benefits of red wine consumption, combined with a Mediterranean diet, were highly publicized over 10 years ago, but were not confirmed in later studies

Alcohol abuse is a worldwide problem, on account of its relatively low cost and widespread availability. Despite the increasing coverage in the popular press of the adverse effects of alcohol consumption, there has been little progress in educating people to consume less alcohol. Alcoholism will diminish

RSC Metallobiology Series No. 1
Mechanisms and Metal Involvement in Neurodegenerative Diseases
Edited by Roberta Ward, David Dexter and Robert Crichton

Published by the Royal Society of Chemistry, www.rsc.org

life expectancy by approximately 4.2 years. Excessive alcohol consumption is detrimental to health; its adverse effects on the liver and gastrointestinal (GI) tract have been well documented. However, the adverse effects evoked by ethanol on the brain have received less attention. Over the past 10 years, the damage evoked by alcohol abuse on the brain has been identified by a range of advanced technologies, such as magnetic resonance imaging (MRI), diffusion tensor imaging (DTI), positron emission tomography (PET) and electrophysiological brain mapping (EEG). MRI and DTI are used together to assess the brains of patients before and after alcohol cessation, while PET enables visualization of the living brain by analysing alcohol effects on the neurotransmitter system as well as on brain cell metabolism and blood flow. EEG records the electrical signals of the brain; alcohol abusers show distinctive electrophysiological profiles, characterized by low amplitude of their P3 components, the presence of which may be a risk marker for alcohol dependence. MR spectroscopy can evaluate the biochemical composition of the brain by observing the brain's principal proton metabolites; low levels of N-acetylaspartate (NAA; the marker of living mature neurons) were abnormally lower in frontal and cerebellar regions and in the frontal white matter region in alcoholics by comparison to controls, indicating neuronal loss (reviewed[6]).

Previous studies also indicated that women may be more susceptible to alcohol toxicity; in other words, female alcohol abusers drink less units of alcohol/day over a shorter period of years to develop alcohol dependence, although the evidence for this remains contradictory. Differences in brain structure were identified; some studies indicated that women showed radiological evidence for greater brain damage from chronic alcoholism than men.[7]

Binge drinking, defined as an intake of at least 5 units over a 2 h period, followed by a period of abstinence, has received considerable attention over the past 5 years. This is primarily because there may be accelerated alcohol-induced brain damage over a much shorter time period, <2–3 y. The alcohol consumption over this time period is often 50% of the weekly recommended intake. Adolescents appear to be the group that indulge most frequently in this type of drinking pattern. Blood alcohol levels between 0.08–0.15 g/100 mL will be reached. Changes in cognitive function have been identified in some individuals, which may be associated with neuroinflammatory changes in specific brain regions, particularly the hippocampus, a region of the adolescent brain where neurogenesis is actively occurring at this stage of their development. In addition, the neurotoxicity associated with binge drinking may also be related to the periods of abstinence between the few days of excessive alcohol consumption, when alcohol withdrawal occurs, which is associated with an increased release of reactive oxygen species (ROS) and reactive nitrogen species (RNS) species together with the excitatory neurotransmitter glutamate. In our recent studies of adolescent binge drinking individuals, compensatory changes occurred in the brain of certain individuals, possibly being indicative of early brain dysfunction.[8] Such patterns of alcohol abuse may increase the individual's vulnerability to develop alcohol-induced brain damage and alcoholism in later life, if such excessive alcohol intake continues.

High intake of alcohol, >5 units/day, will lead to adverse side effects, including dependence, tolerance and craving, which will maintain the alcohol intake at high levels to prevent withdrawal symptoms. During alcohol withdrawal, multiple side effects occur which will accelerate alcohol brain damage. If cessation can be attained, some of this alcohol-induced damage can be reversed. The focus of this chapter will be directed at the brain damage induced by high alcohol intake over a long period of time, >5–10 years. Chronic alcohol abusers during alcohol intoxication exhibit a variety of acute adverse effects, which include ataxia of gait, slurred speech, prolonged reaction times, poor memory consolidation, impaired emotional modulation and compromised judgment. Overall, this will induce impaired judgment, blunted affect, poor insight, social withdrawal, reduced motivation, distractibility attention and impulse-control deficits.

8.2 Definition of the Disease and Clinical Presentation

It is seldom that symptoms of brain damage will be the presenting factor of the chronic alcohol abuser. Alcohol abuse will continue over a long period of time, frequently years, the subject often being in denial that there is a drink problem. Invariably, over the period of alcohol abuse a multitude of social, economic and psychological problems will have developed. In the main, alcohol abusers may only seek clinical help when their lifestyle has descended into chaos, possibly with loss of employment, family and their home. In addition, another clinical symptom may be the presenting factor such that alcohol-induced brain damage is subsequently identified. The *Diagnostic and Statistical Manual of Mental Disorders Edition IV* (DSM-IV-TR) questionnaire (American Psychiatric Association, 2000) is widely used as a determinant of alcohol abuse as well as alcohol-associated cognitive dysfunction and dementia. There are various diagnostic symptoms for alcohol abuse, which include: (a) tolerance; (b) symptoms of withdrawal; (c) excessive consumption; (d) inability to curtail intake; (e) preoccupation with obtaining alcohol; (f) neglect of social, recreational or occupational activities in favour of drinking; and (g) continued alcohol use despite persistent or recurrent physiological and psychosocial problems. The DSM-IV requires three of more of these symptoms for the clinical diagnosis of alcohol abuse. The DSM-IV-TR also includes questions related to memory impairment as well as cognitive disturbances, such as aphasia (inability to use or understand language), apraxia (failure to make purposeful movements), agnosia (difficulty in identifying objects) or disturbance in executive functioning (deficits in planning, organizing, attention and/or changing cognitive strategies). Alcoholic-associated dementia will result in "global" cognitive decline across a wide range of skills involving perceptual motor, visual spatial and abstract problem solving, as well as learning and memory processes. Over 50% of detoxified alcoholics will display some degree of learning and memory impairment.[9] The key factors that are important for brain damage in chronic alcoholics are: (a) the amount of alcohol consumed; a heavy drinking history is a cardinal feature in the criteria for alcoholic brain damage (more than 35 drinks/week for men or 28 drinks/week for women); (b) the length of years of alcohol abuse (in excess of 5 years); and (c) the occurrence of malnourishment. Malnourishment may play an important role in alcohol-induced damage, particularly if there is thiamine deficiency, as seen with Wernicke–Korsakoff syndrome (WKS), where a more adverse cognitive dysfunction occurs with oculomotor disturbances, motor-ataxia abnormalities and global confusion.

Ethanol has a biphasic effect in that low doses are stimulatory while high doses profoundly suppress the nervous system. Brain regions that are implicated in the reinforcing (addictive) and negative properties of alcohol include the ventral tegmental area, the nucleus accumbens, hippocampus, red nucleus of the stria terminalis and, in particular, the central nucleus of the amygdala (Figure 8.1, top).[10] A neuro-anatomic circuit between these different brain structures is involved in the brain arousal–stress systems that produce the negative emotional states that promote the development of addiction (Figure 8.1, top left). The prefrontal cortex will be adversely affected by alcohol abuse in vulnerable individuals. This brain region is involved in attention, rule setting decisions and behavioral control, and executive function, as well as memory, visuospatial processes and motor control (reviewed[11]) (Figure 8.1, bottom).

Over 75% of autopsied brains from chronic alcohol abusers show significant damage. Neuronal loss will occur in specific regions of the cerebral cortex, hypothalamus and cerebellum,[12] but not in the hippocampus. However, there is a decrease in hippocampal volume in human alcoholic brains due to white matter loss. In addition, reduced corpus callosum volume and white matter degeneration occurs in the cerebellum of alcoholics.[13] The mechanism of white matter loss may be myelin loss and the degradation of axonal circuitry.[14] The frontal lobes are more vulnerable to alcohol-related brain damage than other cerebral regions,[15] with decreased neuron density (15–23%) in the frontal cortex of alcoholics.[16] *In vivo* methods have identified frontal cortex abnormalities in alcoholic brains, *e.g.* decreased frontal lobe volume detected by MRI[17] and diminished regional blood flow,[18] as well as decreased amplitude of event-related potentials during the processing of visual targets.[19]

8.3 Genetics of Alcohol-Induced Brain Damage

Early studies indicated that the offspring of families with a history of alcoholism were more likely to become alcoholics,[20] *i.e.* could drink excessive amounts of alcohol before becoming intoxicated. Twin, family and adoption studies confirmed that genetic factors played an important role in the pathogenesis of alcohol dependence, the genetic contribution being more than 50% for an individual to become an alcoholic.[21] Alcohol dependence has been linked to several chromosomal regions and a variety of genes, although progress in identifying individual genes, together with their interaction, has been slow. There are two principal reasons for this lack of success: firstly, the heterogeneity of alcohol dependence, such that classification of alcohol abusers into

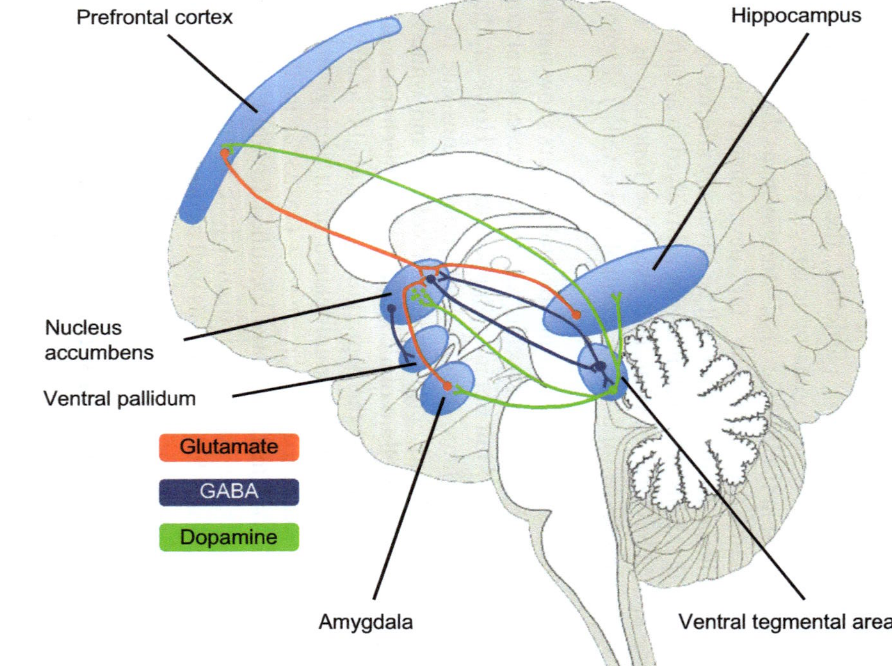

Figure 8.1 *Top*: anatomical picture of the brain regions affected by chronic alcohol abuse. *Bottom*: clinical and pathological features of alcohol-related brain damage.
(Reproduced from Zahr *et al.*[11] with permission from Macmillan).

homogeneous phenotypes, *e.g.* using the Lesch typology,[22] might be a good strategy to solve this problem; secondly, the selection of control subjects is inherently difficult, since many of these subjects may carry the genes for susceptibility to alcohol abuse, but unless they actually consume excessive amounts of alcohol, dependence will not occur. Figure 8.2 shows the various genes that have been implicated in alcohol abuse which, when combined with environmental and mediating factors, contribute to the development of alcoholism. These include monoamine oxidase A (MAOA), catechol-*O*-methyl transferase (COMT) and the genes involved in ethanol metabolism, alcohol dehydrogenase and aldehyde dehydrogenase. Alcohol addiction is likely to be polygenic, with vulnerability arising from the simultaneous impact of functional variations of several genes.

Variants of the alcohol metabolizing genes, alcohol dehydrogenase (*ADH*) and aldehyde dehydrogenase (*ALDH*), cause differential alcohol elimination, which may be important in inducing differences in alcohol metabolism, particularly by increasing levels of acetaldehyde. However, their role in alcohol metabolism in the brain is minimal, catalase possibly playing the major role in its oxidation.[23] However, in early studies, where a standard dose of ethanol was administered orally to controls and alcohol abusers, with variable ADH

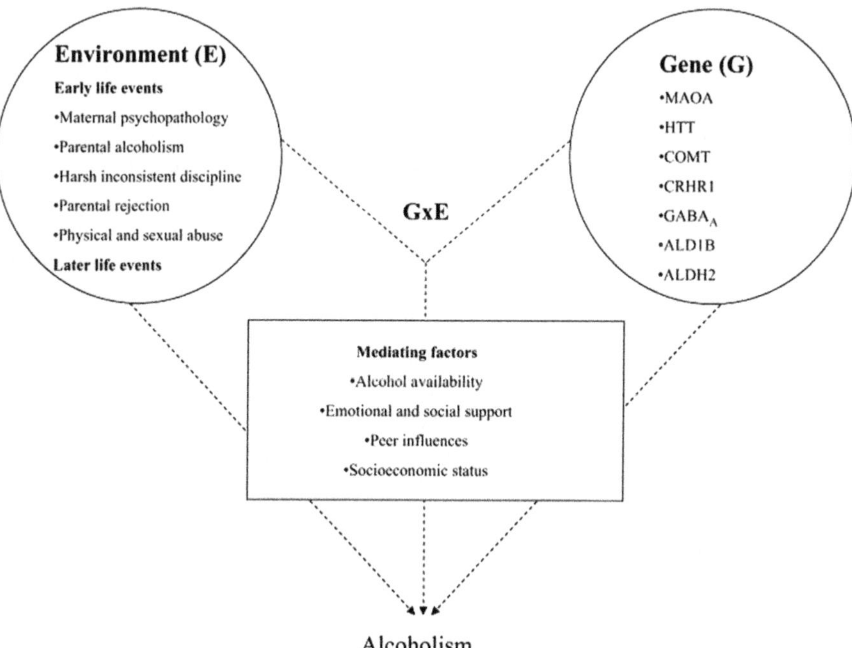

Figure 8.2   Main and interactive effects of genetic and environmental risk factors for alcoholism.
(Reproduced from Ducci and Goldman[79] with permission from Wiley-Blackwell).

genotypes, there were no significant differences in the circulating concentrations of acetaldehyde.[24] For individuals who are homozygous for variants of ALDH2,[25] particularly Oriental individuals, and perhaps ALDH1 variants[26] in Caucasian subjects, the unpleasant side effects elicited by alcohol consumption may act as a protective factor.

The polymorphism exhibited by many of the neurotransmitter systems and their association with excessive alcohol consumption and dependence have been the focus of many studies.[27] The GABAergic system, and pharmacological sensitivity of GABA receptors, play an important role in alcohol tolerance, behavioural and functional neuronal changes, all of which are associated with alcohol dependence. GABA (γ-aminobutyric acid) is the major inhibitory neurotransmitter in the brain. Its receptor $GABA_A$ mediates chloride currents into neurons, which is facilitated by alcohol. Isoforms of the genes encoding for glutamate decarboxylase, *GAD1* and *GAD2* (the rate-limiting enzyme in GABA synthesis), have shown that *GAD1* was associated with initial sensitivity to alcohol. This would relate to the fact that the acute response of the receptor to alcohol is to increase chloride fluxes. In mice a number of ethanol-related behaviours, *i.e.* preference, withdrawal severity and sedation sensitivity, map to four quantitative trait locus regions at which the $GABA_A$ receptor–gene complex are located. Other genes, *e.g.* the *DRD4* gene (dopamine receptor) and the dopamine transporter *SLC6A3* (which influences impulsivity), monoamine oxidase A (which metabolizes dopamine), and 5-hydroxytryptamine serotonin receptor 1B (which is important in aggression and alcohol preference), all show polymorphisms although, as yet, no links with alcohol abuse have been identified.

In animal studies a transition of A silent G–A at nucleotide 603 in exon 5 of the excitatory amino acid transporters, the *EAAT2* gene, is associated with risk-taking behaviour in alcohol abuse, while mutation of the Period gene (*PER^{Brdm1}*) results in increased alcohol intake.[28] Both of these changes will result in alterations in glutamatergic transmission. The expression of the *N*-methyl-D-aspartate glutamate receptor (NMDA) complex is under multifactorial control. There were statistically significant associations of alcohol dependence with genetic variation in the NMDA-dependent AMPA receptor trafficking cascade.[29] Real-time RT-PCR of RNA isolated from the superior frontal (SFC) and primary motor (PMC) cortex tissue of chronic alcoholic subjects +/− liver cirrhosis showed a significantly lower level of expression of NR1, NR2A and NR2B subunit mRNA in cirrhotic patients,[30,31] although this may be alcohol induced.

The dopamine receptor gene *DRD2* has received considerable attention, where a polymorphism, Taq1-A1, was originally identified. This was localized ∼ 10 kb downstream of *DRD2* in a neighbouring gene, *ANKK1*. However, both positive and negative associations have been reported. Other dopamine receptors have also been investigated for their possible association with alcohol abuse and dependency, although no positive results have been reported.

Particular attention has been focused on genes involved in the immune system since alterations in their expression could ultimately contribute to brain damage, alcohol dependence, depression, anxiety and cognitive impairment. Polymorphisms in the nuclear factor of the κ-light polypeptide gene enhancer in B-cells 1/the p50 precursor gene (NFκB1) are highly correlated with an increased risk of alcoholism: a direct link between NFκB p50 and alcohol dependence has been suggested. Specific TNFα and IL10 alleles, as well as polymorphisms in the *IL-1* gene, have been associated with alcohol abuse.[32]

Epigenetic alterations to the genome, *e.g.* DNA methylation and histone modifications, are important mechanisms underlying addiction and the neurobiological response to addictive substances (Figure 8.3). Various studies have demonstrated the important role of epigenetic modifications in regulating behavioural responses to alcohol exposure. For example, nuclei isolated from microglia present in the brains of chronically alcohol-fed rats showed both chromatin conformation into a condensed state as well as alterations in the non-histone nuclear protein composition.[33] Such changes may influence the microglia response to ethanol in the brain.

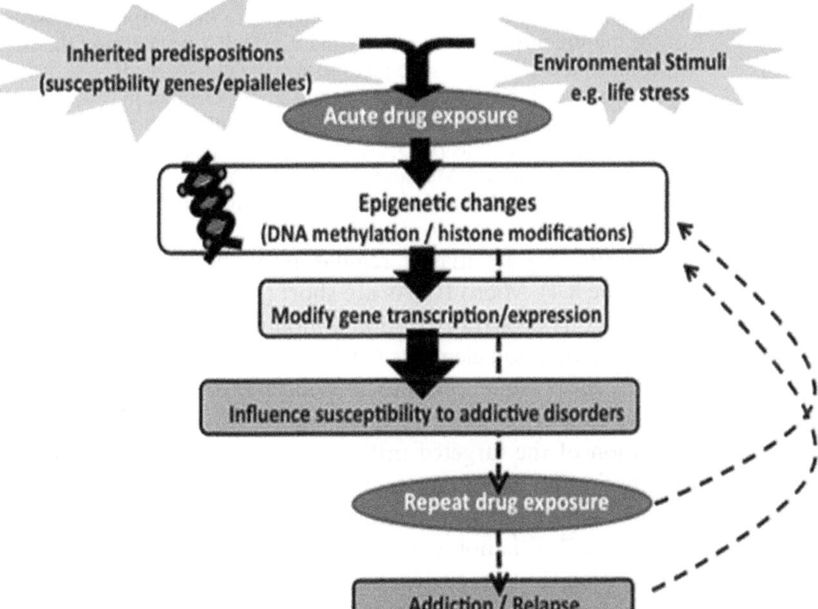

Figure 8.3 The proposed relationship between inherited predispositions, environmental factors, exposure to addictive substances and vulnerability to addictive disorders. When exposed to adverse environmental stimuli, individuals carrying susceptibility genes or epialleles predisposing to addictive behaviour may have an increased risk of developing addiction. Acute drug use may produce enduring alterations in gene expression *via* epigenetic changes that influence susceptibility to addictive disorders. Enhanced vulnerability to drugs of abuse will then feed back into increased risk of future drug use (as shown by the *dashed arrows*) that bring about further modifications to the epigenome and gene expression. (Reproduced from Wong *et al.*[80] with permission from Wiley-Blackwell).

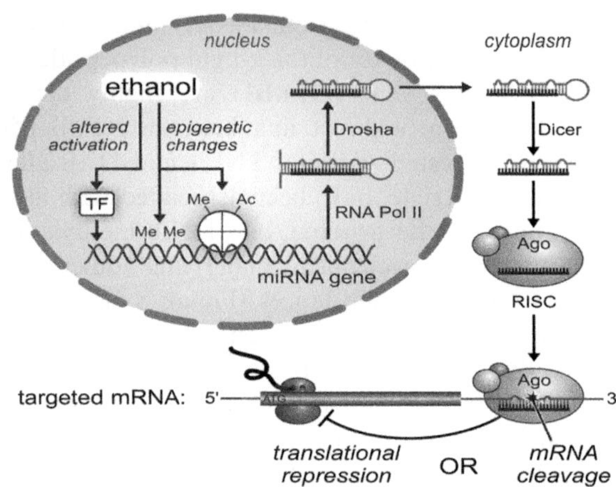

Figure 8.4   Micro RNA biogenesis and potential influences of alcohol. Hairpin primary miRNA transcripts are successively cleaved by RNases Drosha and Dicer while undergoing transport from the nucleus to the cytoplasm. Association of the mature miRNA with an Argonaut protein (Ago) directs the complex to complementary target sequences in specific messenger RNAs, leading to their cleavage or disruption of their translation. Ethanol may affect miRNA expression by altering activation of transcription factors (TF) and/or epigenetic modifications of DNA and DNA-associated histone complexes, including methylation (Me) and acetylation (Ac)
(Reproduced from Miranda *et al.*[35] with permission from Wiley).

Micro RNAs (miRNAs) are master regulators of the cellular transcriptome and proteome (Figure 8.4). Micro RNAs are short (17–24 nucleotides long) non-coding RNAs and act as post-transcriptional modulators of gene expression by binding to miRNA-recognition elements (MREs) in their target genes. The target genes are defined by short sequences in their 3′-untranslated regions (UTRs) that are complementary to a given miRNA. This will result in suppression of translation or degradation of the targeted mRNA transcript, or both. miRNAs are abundant in the brain and play important roles in various biological processes, including neuronal differentiation, synapse formation and plasticity, as well as neurodegeneration.[34] Ethanol may affect miRNA expression by altering transcription activation and/or epigenetic modifications of DNA and DNA-associated histone complexes. A study of the prefrontal cortex from post-mortem alcoholic brains[34] showed that approximately 35 human miRNAs were upregulated in these samples, which included miRNAs that target cell death-related genes and multiple immune function genes. In addition, the genes involved in calcium- and voltage-dependent potassium channels and myelin gene expression are targets for miRNAs after chronic alcohol consumption.[35]

8.4   Proteins Involved in Alcohol-Induced Brain Damage

Chronic alcohol abuse will induce damage in brain regions that are responsible for neurocognitive function. Such damage could be permanent or may be reversible after detoxification. Using high-throughput proteomics approaches and two-dimensional electrophoresis (2-DE)-based proteomics, Matsumato[36] studied post-mortem brains from two groups of alcohol abusers: (a) uncomplicated alcohol abusers and (b) alcohol abusers with hepatic cirrhosis, who were matched for age and gender, as well as the period of time before post-mortem intervention and brain pH. All alcoholic abusers fulfilled the DSM IV criteria. A wide number of proteins were differentially expressed in a number of brain regions in both groups of alcohol abusers, with abnormalities in vitamin B1 (thiamine)-related biochemical pathways, *i.e.* thiamine-dependent proteins, transketolase and pyruvate dehydrogenase E1β-subunit, in the dorsolateral prefrontal cortex (both grey and white matter), genu (a frontal part of the corpus callosum) and cerebellar vermis. Furthermore, Matsumato[36] suggested that decreased thiamine status might be an important determinant of cognitive impairment in non-WKS alcohol abusers. A large number of metabolism-related enzymes, *e.g.* glycolysis and the tricarboxylic acid cycle, were also found to be differentially regulated. Two enzymes of the non-oxidative branch of the pentose–phosphate pathway (PPP), thiamine-dependent transketolase and transaldolase, were altered in both alcoholic groups.[37,38] Three thiamine-dependent enzymes, transketolase, dihydrolipoamide dehydrogenase and pyruvate dehydrogenase E1 β-subunit, showed altered expression in the cerebellar vermis of both alcohol groups. A number of previous studies have reported glucose hypometabolism in the frontal regions of chronic alcoholics.[39–41] In the hippocampus there were significant changes in glutamine synthetase expression, possibly due to the excessive release of ammonia, which might induce brain damage in this region. This enzyme is predominantly located in astrocytes[42] and plays an important role in protecting neurons from toxicity induced by glutamate and ammonia by converting glutamate into glutamine, thereby taking up excess ammonia.[42]

Wistar rats exposed to alcohol *via* the vapour method for 14 h in every 24 h for 26 weeks showed a range of proteins which were up- or down-regulated in prefrontal cortex, dorsal striatum, corpus callosum genu, CC body anterior vermis and the anterior dorsal lateral cerebellum.[43] The functions of these proteins were mainly associated with neurotransmitter systems in the prefrontal cortex and cell metabolism in the dorsal striatum. In addition, stress proteins were increased only in the prefrontal cortex, the anterior vermis and the anterior dorsal lateral cerebellum. Clearly, a range of proteins are altered in the brain after chronic alcohol abuse, but their exact relevance to the induced brain damage awaits clarification.

Acetaldehyde may react with the ε-amino group of lysine or the α-group of N-terminal amino acids, as well as with dopamine to form the tetrahydro-isoquinoline salsolinol (Figure 8.5). Increased levels of salsolinol in various brain regions after alcohol abuse, *e.g.* nucleus accumbens, caudate putamen and mid-brain, may play a role in the behavioural and reinforcing effect of alcohol. In a recent study it was shown that the core and shell regions of the nucleus accumbens showed different responses to salsolinol which may be due to their receptor expressions, particularly of the opioid receptor.[44]

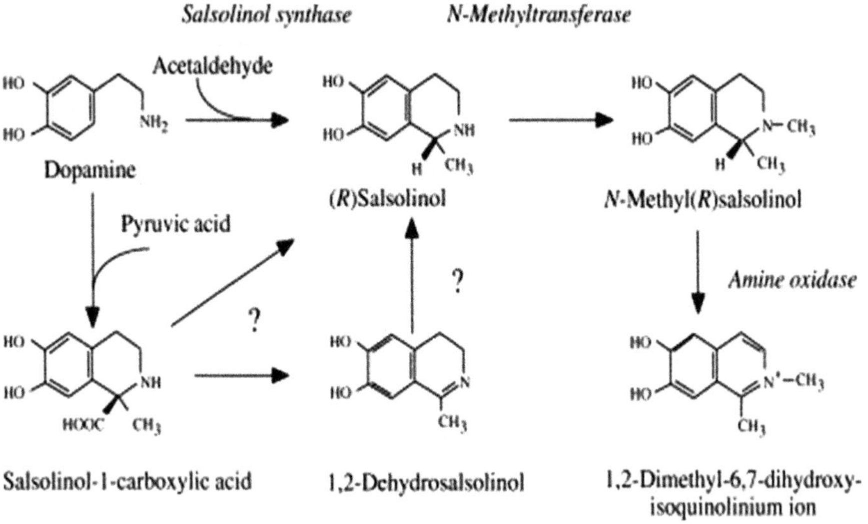

Figure 8.5   Biosynthesis pathway of (R)-salsolinol derivatives in human brain. The enzymatic condensation of dopamine with acetaldehyde or pyruvic acid is catalyzed by (R)-salsolinol synthase to yield (R)-Sal and (R)-Sal-1-carboxylic acid. However, the enantioselective synthesis of (R)-Sal from (R)-Sal-1-carboxylic acid and 1,2-dehydrosalsolinol was not confirmed. N-Methyltransferase catalyzes the N-methylation of (R)-Sal into N-methyl-(R)-Sal, which is further oxidized to the 1,2-dimethyl-6,7-dihydroxyisoquinolinium ion (DMDHIQ$^+$).
(Reproduced from Naoi *et al.*[81] with permission from Elsevier).

8.5   Neuroinflammation

Chronic inflammation is commonly associated with alcohol-related medical conditions, thereby acting as an etiological factor in the initiation and progression of adverse effects. Indeed, alcohol abusers with alcoholic liver disease may show a high level of circulating proinflammatory cytokines,[45] as a result of alcohol metabolism leading to ROS production and hypoxia, as well as from the release of lipopolysaccharides from gut microflora (Figure 8.6). Such systemic inflammation may be a factor in inducing neuroinflammation in the brain.[46]

Ethanol has a modulating effect on immunity, inducing neuro-inflammation, particularly when other stimulatory molecules are present, which may be an important factor in the etiology of alcoholic brain damage. Altered innate and adaptive immune function occurs in chronic alcohol abusers; in fact, they are immune deficient. Such patients are more prone to infections, have a diminished ability to fight infection as well as an increased risk of developing cancers, particularly of the upper respiratory and gastrointestinal tract. Many studies have shown that low-grade infection will be exacerbated in the presence of alcohol. For example, when chronic alcohol ingestion is combined with

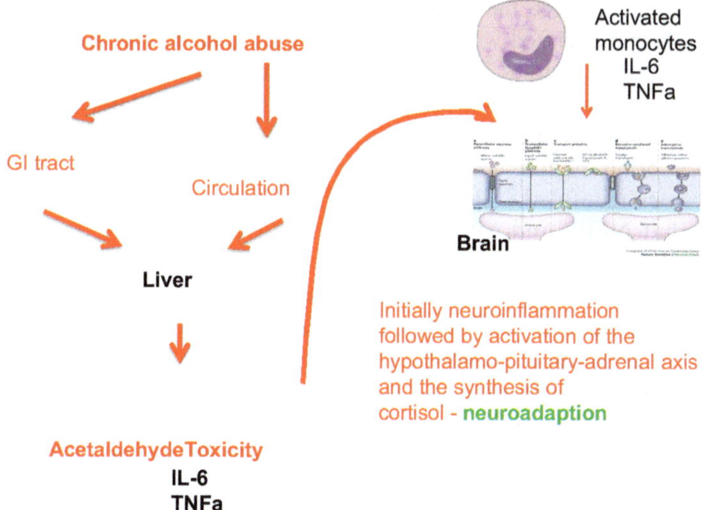

Figure 8.6 Release of pro-inflammatory cytokines from the liver. After alcohol ingestion, ethanol will be absorbed directly into circulation or *via* the gastrointestinal tract to reach the liver where it will be principally metabolised by alcohol dehydrogenase and aldehyde dehydrogenase. The acetaldehyde released may have an adverse effect on hepatic cells, inducing inflammation and the release of pro-inflammatory cytokines. These may directly cross the blood–brain barrier to evoke inflammation in the brain *via* microglia activation. In addition, monocytes in the blood may be activated, which are able to cross the blood–brain barrier and further induce an inflammatory response by microglia. Such an inflammatory milieu may induce neuronal damage. With time, neuroadaption may occur where certain anti-inflammatory agents are released, *e.g.* cortisol, to counteract the pro-inflammatory state.

specific agonists for Toll-like receptor 3, TLR3 [polyinosine-polycytidylic acid (poly:I:C) sensitized with D-GaIN],[47] or TLR4 (lipopolysaccharide, LPS),[48] a robust inflammatory response occurs with the release of proinflammatory cytokines (*e.g.* IL1β and TNFα). Overall, chronic ethanol alone has only a small priming effect on cells, but does not initiate an intense inflammatory response in glial cells.

Alterations in the gene expression of various transcription factors, *e.g.* cAMP-response element-binding protein and NFκB, will contribute to the neuro-adaptation induced by chronic alcohol consumption. In the brain, NFκB regulates a variety of pathways which include synaptic plasticity (which is important in underlying memory and learning formation), long-term potentiation and depression in addition to neuro-inflammation and neuronal survival.[32] Inducible NFκB is present in synapses, while glutamatergic stimulation activates retrograde transport of p65 protein from the synapse to the cell nucleus, thereby indicating that NFκB is sensitive to changes in gene expression. Since both IκBα and activated IKK are present within the axon initial segment (where the action potential is generated), this would indicate that NFκB is associated with the processing of neuronal information.

In our earlier studies,[49] we reported that NFκB activation was down-regulated in the cortex of chronically alcoholized rats. In later studies of post-mortem alcoholic brains, NFκB was also shown to be down-regulated in the precortex brain region. The factors involved in this down-regulation have recently been identified.[50] Post-mortem brains from 15 chronic alcohol abusers showed that NFκB and p50 homodimer DNA binding were down-regulated, while levels of p65 (RelA) mRNA were attenuated in the prefrontal cortex compared to controls. This would indicate an adaptive mechanism to protect the remaining neurons against neurotoxicity and neuroinflammation.

Neuroinflammation is involved in the pathogenesis of alcohol-related brain damage. Microglia can exist in multiple states, notably a pro- (M1) or anti-inflammatory state (M2). Activated microglia are present in post-mortem alcoholic brains, particularly in the cingulated cortex, which was confirmed by the presence of two microglia marker proteins, glucose transporter type 5 and ionizing calcium binding adaptor protein, Iba1. In addition, a proinflammatory chemokine, monocyte chemotactic protein (MCP-1), was present in the ventral tegmental area, substantia nigra, hippocampus and amygdala of alcoholic post-mortem brains,[51] which was possibly released from the activated microglia. Such elevation of MCP-1 could directly induce neuronal apoptosis as well as enhancing the microglia synthesis of more proinflammatory cytokines. However, no differences in the spectrum of microglial morphologies were evident between the alcoholic or control post-mortem brains. Further studies are clearly warranted.

Many studies have been reported on the effect of various stimulants on glial cells isolated from animal models of chronic alcoholism or of primary or immortalized glial cell cultures after incubation with ethanol. For the latter studies, mixed glial cultures are more representative of the *in vivo* situation than microglia, astrocytes or oligodentrocytes alone. Intracellular signal trans-duction will be cell-type specific. Therefore the overall effect of ethanol on the inflammatory activation of pure microglia, astrocytes and mixed glial cultures may be similar, but the specific effects of ethanol on intracellular signal transduction may differ (Figure 8.7).

8.5.1   Astrocytes and Alcohol

Prolonged ethanol consumption decreases astrocyte numbers, as exemplified by changes of glial fibrillary acidic protein (GFAP) (marker protein for astrocytes) in the dorsolateral and orbitofrontal cortex of human post-mortem brains,[52] as well as in a strain of alcohol preferring rats, where a lower number of astrocytes was present in the prefrontal cortex.[52] Astrocytes can detect changes in the molecules released from neurons during transmission and respond accordingly, by either regulating the extracellular medium or releasing various factors to alter the activity of neurons or other connecting glial cells. Astrocytes also have membrane receptors for a variety of neurotransmitters, *i.e.* glutamate,

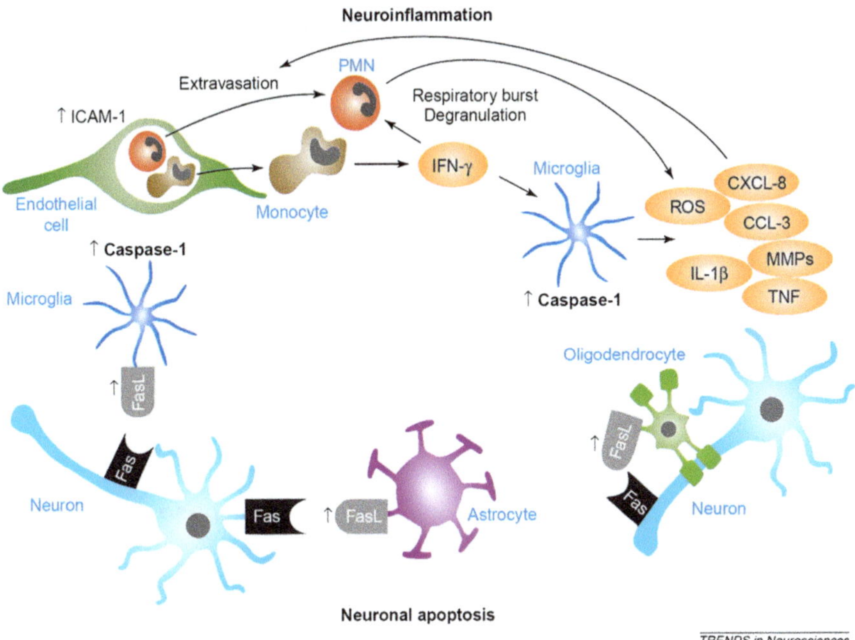

Figure 8.7 Pathway of neuroinflammation induced by ethanol. Interaction between astrocytes, microglia and oligodentrocytes.

dopamine, norepinephrine, serotonin, GABA, *etc.*, such that neuronal activity can be controlled. Alcohol abuse may induce changes in the function of astrocytes by altering the release of brain-derived neurotrophic factor (BDNF) (which protects the neurons from different insults), decreasing glucose uptake (due to a decrease in glucose transporter protein expression, GLUT1)[53] or increasing monosaccharide uptake and GLUT1 protein levels.[54] Astrocytes play an important role in the reuptake of glutamate released by neurons during synaptic activity, *via* transporters EAAT-2 and EAAT1, such that any excesses of glutamate remaining in the synaptic cleft could induce toxicity. Alterations of such glutamate transport may be related to genetically inherited vulnerability to alcoholism.

Activation of Src kinase (Figure 8.8) by ethanol, *via* the interactive tyrosine phosphorylation of TLR4, will lead to a series of de-phosphorylation and phosphorylation events to induce the activation of cytosolic phospholipase A2, and synthesis of arachidonic acid. This will be metabolized by cyclooxygenase 2 to prostaglandin E2.[55] In addition, after chronic ethanol treatment, iNOS, IL-1β, IRAK, ERK1/2 p38 and JNK are released from astrocytes isolated from the cortex of chronically fed animals, which could induce stress and apoptotic cell death.[56] Ethanol may target phospholipase D in astrocytes, causing a decrease in phosphatidic acid, decreased activation of atypical protein kinase C zeta and the decreased downstream activation of p70S6 kinase and NFκB.

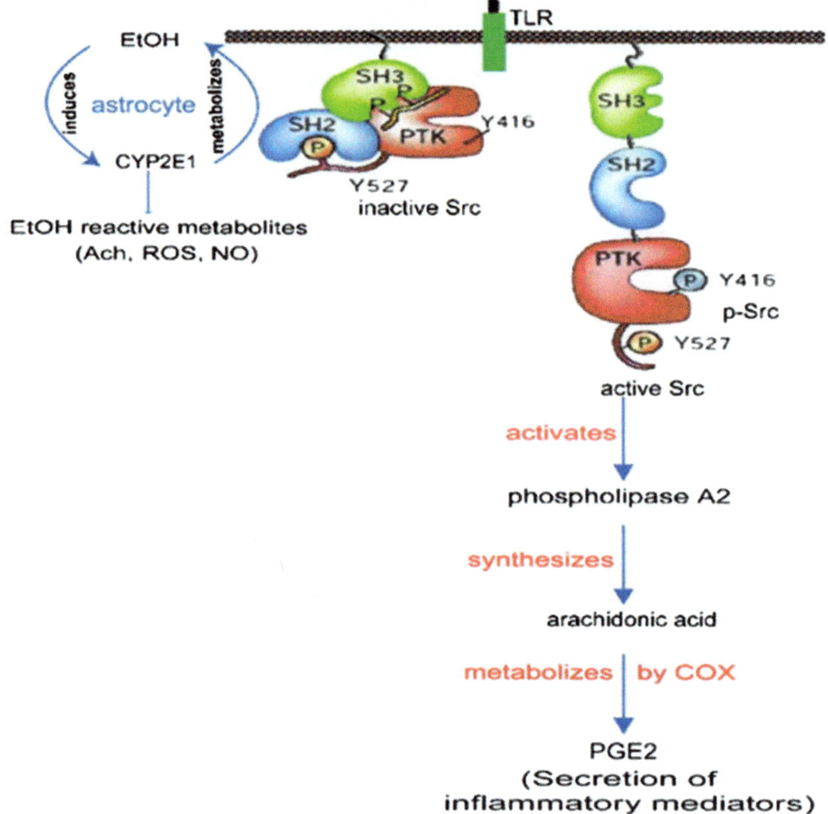

Figure 8.8   The Src family of tyrosine kinases are cytosolic kinases which are normally thought to be regulated through tyrosine kinase receptor mechanisms. The receptor ligand induces dimerization of the receptors. The receptor dimers then cross-phosphorylate each other (*light blue circle*) to give rise to the activated receptor complex. The receptor complex then acts as a platform for the phosphorylation (on tyrosines) of other cellular proteins. Among the target proteins is the Src family of tyrosine kinases. (reproduced by permission from Springer)

8.5.2   Microglia and Alcohol

Microglia are not directly involved in the regulation of neurotransmission and nerve impulse conditions. However, they may influence the survival of neurons indirectly as they are able to release trophic factors as well as expressing receptors for neuropeptides and neurotransmitters. Microglia function is altered by long-term ethanol exposure, as exemplified by decreased phagocytic function, altered superoxide anion production in resting cells, and reduced NO release after *ex vivo* stimulation, thereby altering the balance between NO and $O_2\cdot^-$. This reduction in NO release may be caused by ethanol's inhibition of the phospholipase D–tyrosine kinase pathway which is involved in

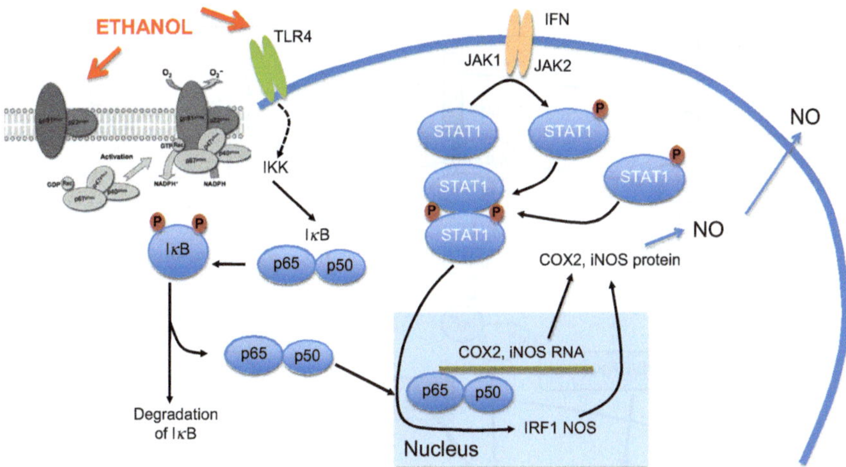

Figure 8.9 Synthesis of NO and O_2^- by stimulated microglia. The release of nitric oxide, NO and superoxide, O_2^-, by stimulated microglia are key signalling events in alcohol induced brain damage. Alcohol will activate Toll-like-Receptors, TLR4, to trigger a signalling cascade which results in the release of pro-inflammatory cytokines, such as TNFα, IL1β and transcription factors. There are numerous potential binding sites for transcription factors on the NOS promoter region which will be activated and initiate NO synthesis. NADPH oxidase (nicotinamide adenine dinucleotide phosphate-oxidase) is a membrane-bound enzyme complex. NADPH oxidase generates superoxide by transferring "electrons from NADPH inside the cell across the membrane" and coupling these to molecular oxygen to produce superoxide.

the phosphorylation, as well as the activation of NADPH oxidase and myeloperoxidase. The decrease in reactive oxygen intermediate formation, tyrosine kinase-induced phosphorylation, and activation of transcription factors NFκB will lead to a decrease in expression of iNOS mRNA.[57] Such alcohol-induced immunosuppression or immunomodulation is well recognized in alcoholic subjects, making them more susceptible to opportunistic infections. Activated microglia release cytokines, chemokines, proteases and reactive oxygen or nitrogen species, which function in a synergistic and/or antagonistic manner, eventually leading to neurodegeneration (Figure 8.9). The activation of microglia is regulated by neuronal signals as well as astrocytes and various systemic signals.

NFκB serves as a common downstream signalling component of all known TLR signalling pathways. Ethanol selectively inhibits LPS-induced microglia activation, with reduced NO release and IL-1β expression, but does not alter the IFNγ responses. It was suggested that ethanol inhibits NFκB signalling somewhere between DNA binding of NFκB and the transactivation of target genes (Figure 8.10). The site of ethanol action was suggested to be p300, which regulates p65 transactivation.[33] More recently, studies of post-mortem brains from 15 chronic alcohol abusers showed that NFκB and p50 homodimer as

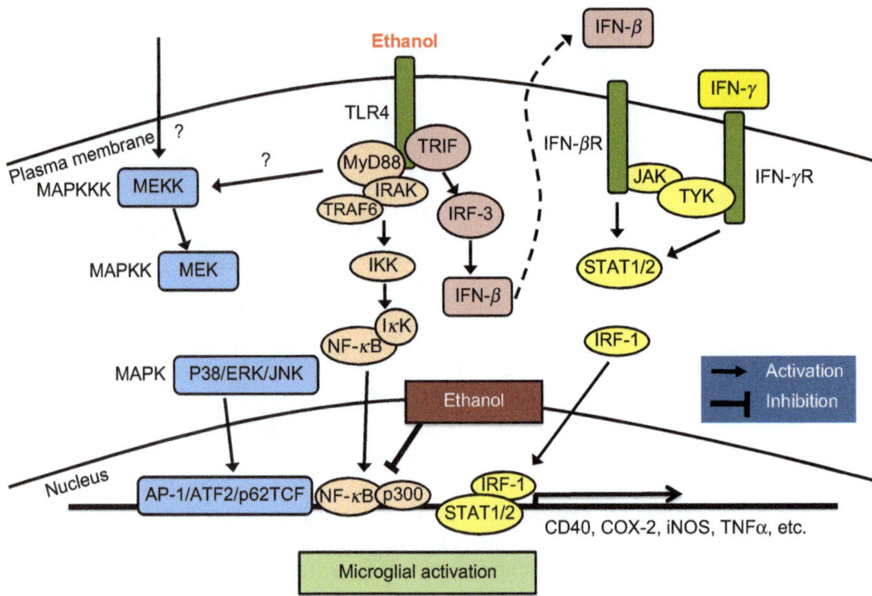

Figure 8.10   Schematic diagram of microglial signal transduction. TLR4 signalling is initiated with NFκB activation and IFNβ production after ethanol stimulation. TLR4, Toll-like receptor 4; MAPK, mitogen-activated protein kinases.
Adapted from ref. 33.

well as RELA gene expression are down-regulated in the prefrontal cortex by comparison to controls. Such down-regulation could result from a decrease in RelA gene expression as well as alterations in stoichiometry of the p65, p50 and IKKβ proteins. This would indicate an adaptive mechanism to protect the remaining neurons against neurotoxicity and neuroinflammation.[32,50]

8.5.3   Toll-like Receptors (TLRs)

Ethanol will increase the numbers of TLR3 and TLR4 receptors which will activate inflammatory pathways by a variety of transcription factors, *e.g.* AP1 and NFκB, *via* MAPKs and MYD88-independent pathways to stimulate neuronal death (Figure 8.11).[58] TLR4 receptors are present on the cellular membranes of both astrocytes and microglia and will be activated by ethanol.[59] It is unclear how alcohol activates TLR4 and its downstream signalling pathways. It has been hypothesized that ROS (released from ethanol meta-bolism by the cytochrome P450 2E1, CYP2E1, inducible cytochrome or catalase) induces the dissociation of the inactive Src complex and the phos-phorylation of tyrosine 416 in PTK, resulting in phosphorylation and formation of active Src, which activates cytosolic phospholipase A2 and leads to the release of cyclooxygenase (phosphorylate) and other proinflammatory agents.[55]

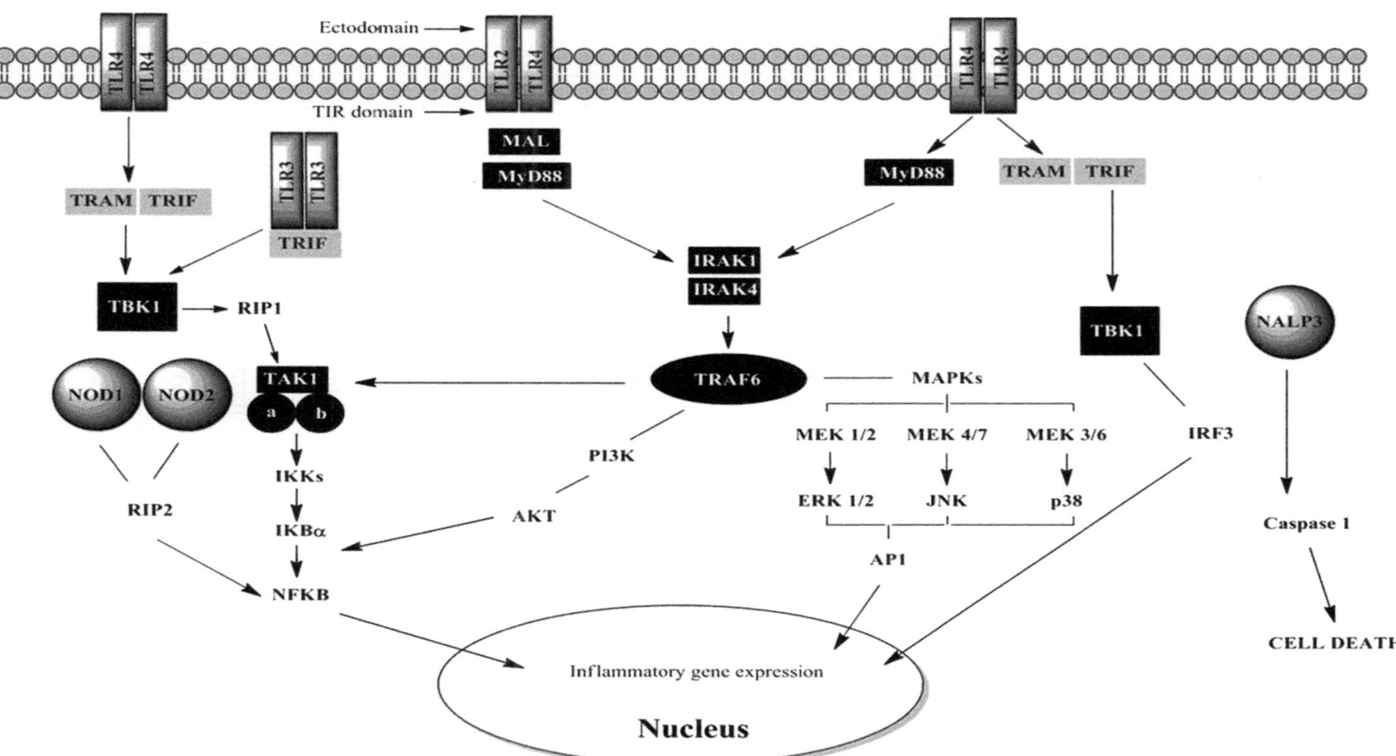

Figure 8.11   Activation of Toll-like receptors activates various pathways to produce the release of proinflammatory cytokines.

8.5.4 Oligodendrocytes and Alcohol

Oligodendrocytes are extremely sensitive to ethanol exposure, with increased myelin degeneration occurring.[52] Chronic ethanol administration to rats induced down-regulation of proteins involved in myelination, with immuno-chemistry showing altered myelin morphology, reduced numbers of myelin basic protein positive fibres and oligodentrocyte death. Furthermore, 41–47% of axons showed myelin sheath disarrangement in the cerebral cortex and corpus callosum.[60] Microglia activation and the release of proinflammatory cytokines, together with reactive oxygen and nitrogen species, particularly $ONOO^-$, and myeloperoxidase activity may have adverse effects on oligodendrocyte function.[61]

8.5.5 Alcohol and Mitochondria

Mitochondria in the liver play a major role in alcohol metabolism due to the fact that ALDH oxidizes acetaldehyde to acetate with the generation of high amounts of NADH. In contrast, the K_m of the ADH isoform (ADH5) and of the various isoforms of ALDH found in the brain are extremely high, implying that further metabolism of alcohol by these enzymes is improbable. However, local metabolism of ethanol by CYP2E1 and catalase could increase both acetaldehyde and ROS in brain cells, thereby adversely affecting the mito-chondrial respiratory pathway. This could result in defects of complexes I, III, IV and V, together with loss of mitochondria function. The oxidation of mitochondrial protein is a recognized feature of chronic alcohol abuse. Coleman and Cunningham[62] established a key link between the chronic alcohol-related defects in complexes I, III, IV and V in the liver and losses in the 13 mitochondrial encoded polypeptides and redox centres that comprises the complexes of the oxidative phosphorylation system. Furthermore, proteomic analysis revealed that 40 additional mitochondrial proteins had altered levels in response to chronic alcohol consumption, including several key enzymes involved in energy metabolism, such as β-oxidation, the trichloroacetic acid cycle and amino acid metabolism.[63] Increased amounts of peroxynitrite could also be produced, thereby enhancing mitochondrial toxicity. Brain mito-chondria are the major target of the oxidative stress generated during ethanol intoxication and withdrawal. In mice it has been shown that such stress will initiate degradation and induce damage to mtDNA in mice. During alcohol withdrawal the increased release of glutamate will increase intracellular Ca^{2+} and ROS, both of which will open mitochondrial permeability transition pores, thereby allowing the passage of solutes and causing mitochondrial swelling and decreased efficacy of the mitochondria.[63]

8.5.6 Alcoholic Brain Damage and Oxidative Stress

The association between alcohol abuse and excessive free radical production in many different organs has been well documented. Chronic alcohol abuse induces the release of both ROS and RNS in specific cell types, *i.e.* neurons, astrocytes and microglia. In addition, there may be leakage of $O_2 \cdot^-$ from ethanol-damaged mitochondria. Such free radicals will be particularly damaging since in the brain there is a diminished ability to protect itself from these species, as exemplified by low activities of cytoprotective enzymes and of reduced glutathione.[64] As already described, ROS are generated during ethanol metabolism in the brain by cytochrome P450, CYP2EI (induced by prolonged alcohol abuse), as well as by catalase. Therefore in brain regions where cytochrome CYP2EI is induced after chronic alcohol abuse, there may be an increased vulnerability to oxidative stress. Since CYP2EI expression is increased in the olfactory bulbs, frontal cortex, hippocampus and cerebellum in male rats administered ethanol, 3 g/kg for 7 days,[65] oxidative damage may be greater in these regions. High amounts of inducible CYP2E1 were present in the superior frontal cortex and hippocampus in conjunction with a reduction in the number of cerebellar Purkinje cells, after chronic alcohol intake in rats, which was associated with cortical neuronal loss.[66]

Reactive nitrogen species, particularly NO and peroxynitrite, $ONOO^-$, are likely to be toxic to brain cells since they can rapidly transverse cellular membranes to interact with various cellular constituents, *e.g.* peroxynitrite released from activated microglia will nitrate protein tyrosine residues to form 3-nitrotyrosine, particularly in frontal cortical neurons after chronic ethanol administration. Chronic ethanol abuse has a differing effect on the different isoforms of iNOS. iNOS is inhibited after chronic ethanol abuse, due to inhibition of NFκB and the inhibition of signal transducer and activator of transcription-1 (STAT-1) activation.[67] Other isoforms of NOS are also altered by chronic ethanol abuse.[68]

Few studies have been directed towards the detection of 4-HNE, which is now well established as a reliable marker of oxidative stress, in brains of animals or humans after chronic alcohol abuse. In one study of brain homogenates, increases of 4-HNE were identified in mice which had received the chronic ethanol liquid diet-fed regime,[69] while the selective activation of NADPH oxidase (NOX) by chronic alcohol administration enhanced ROS and 4-HNE, predominantly in astrocytes and microglia.[69]

8.6 Metals Involved in Alcoholic Brain Damage

As yet, a role for metals, such as iron, copper and zinc, in the exacerbation of alcoholic brain damage has not been identified.

8.7 Neurotransmitters Involved

Chronic alcohol abuse will alter the function of many neurotransmitter receptors which are involved in the inhibitory and excitatory systems in the CNS. These include GABA, glycine, glutamate, norepinephrine, dopamine, serotonin, acetylcholine, cannabinoid and nicotine, as well as a range of transporters for adenosine, norepinephrine, dopamine and serotonin.

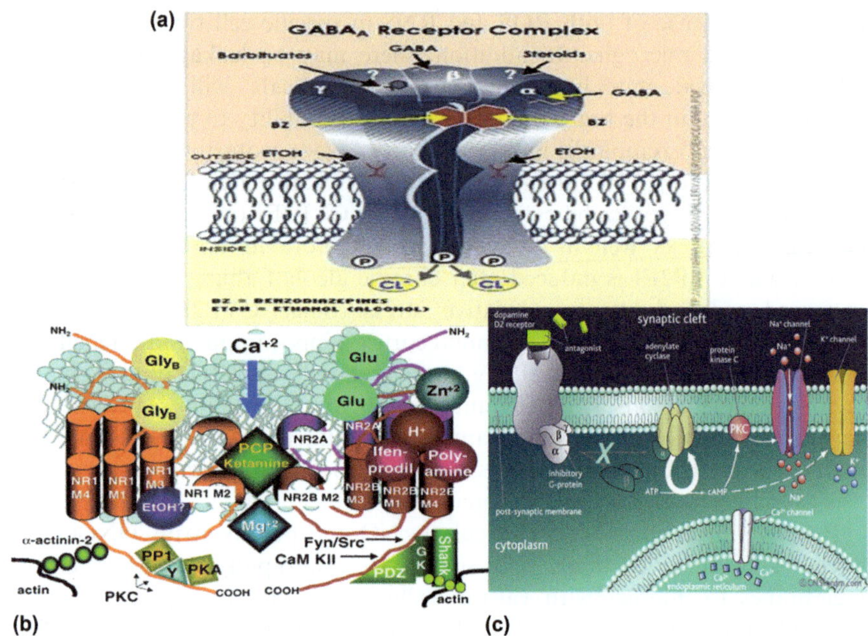

Figure 8.12 (a) GABA receptor; (b) NMDA receptor; (c) Dopamine receptor. (Reproduced from Rang *et al.*[78] with permission from Harcourt).

GABA receptors will mediate the sedative effects of alcohol, motor inco-ordination, tolerance and dependence (Figure 8.12a). Long-term alcohol exposure will decrease $GABA_A$ receptor function, thereby decreasing sensitivity to neurotransmission. The functional $GABA_A$ receptor comprises five subunits, normally two α, two β and one γ subunits. The combination of different sub-compositions will induce different functional properties.

N-Methyl-D-aspartate glutamate receptors (NMDAR) are multimeric transmembrane proteins which consist of several different types of subunits (Figure 8.12b). Many NMDAR subunits have been identified in neurons, including the ubiquitously expressed NR1 subunit (which occurs in eight distinct isoforms), NR2 (A, B, C and D) and NR3 (A, B) subunits. The NR1 subunits are required to form an active ion channel, while incorporation of the various NR2 subunits regulates the receptor channel activity by altering the channel kinetics and mediating the pharmacological effects of alcohol. It is proposed that NMDAR exists as tetrameric complexes made up of at least one NR1 and one NR2 subunits.[70] In addition, the NR2 subunits are phosphorylated by the Src family protein tyrosine kinases (PTKs), which will induce up-regulation of channel function. Various co-factors such as Mg^{2+}, Zn^{2+} and glycine also modulate NMDA function. Ethanol will target the NMDA receptor and induce a variety of adverse effects, such as tolerance, dependence, withdrawal, craving and relapse. Ethanol potently inhibits NMDARs, but with prolonged ethanol exposition there is a compensatory "up-regulation" of their functions. Interestingly, ethanol does not affect NMDAR function in all brain regions in an identical fashion, which could reflect the different composition of the subunits present.

The mesolimbic dopamine system (Figure 8.12c) is a key player in the development and reinforcement of alcohol abuse. Initially, alcohol will enhance the activity of the mesolimbic dopamine system, thereby enhancing extra-cellular dopamine release, but with time as more alcohol is consumed it will be necessary to consume higher quantities to elicit the enhanced dopamine release.

8.8 Therapeutics

The obvious treatment for alcohol abuse is the cessation of alcohol consumption. This is difficult to achieve in many alcohol abusers since craving is an inherent problem. Currently there are three drugs available for the treatment of alcohol dependence: disulfiram (Antabuse®), naltrexone (ReVia®) and acamprosate (Campral®). Because of the wide heterogeneity in alcohol abusers, the use of such drugs may not be efficacious in every individual.

Disulfiram (Figure 8.13a) is an aversion-based psychopharmacotherapy. Its mode of action is to inhibit mitochondrial aldehyde dehydrogenase. If alcohol is ingested while the patient is taking disulfiram, then flushing of the skin, accelerated heart rate, shortness of breath, nausea, vomiting, *etc.*, will occur within 5–10 minutes, due to the increase in circulating acetaldehyde. This major metabolite of alcohol will increase 4- to 5-fold compared to that of the normal acetaldehyde concentration after alcohol ingestion, *i.e.* from $4 \mu M$ to $20 \mu M$. The major problem with the use of the drug is patient compliance, since the action of disulfiram is reversible and the patients can cease its administration if the craving for alcohol is too great. There are also reports that it is also a potent metal ion chelator and will inhibit dopamine-β hydroxylase, an enzyme which converts dopamine to norepinephrine, an important neurotransmitter in both the CNS and autonomic nervous system (reviewed[71]).

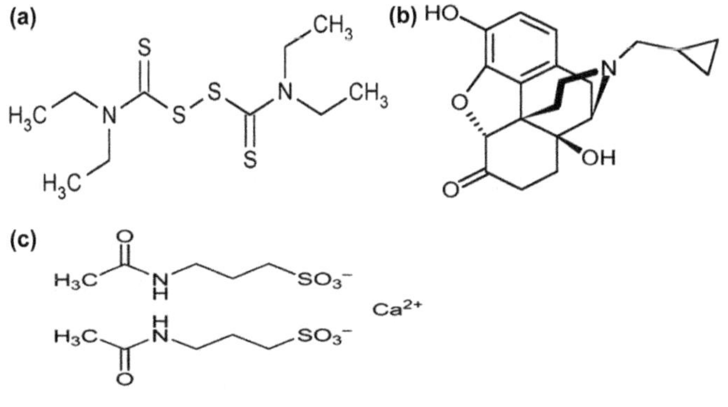

Figure 8.13 Drugs used for the treatment of alcohol abuse: (a) disulfiram; (b) naltrexone; (c) acamprosate.

Naltrexone (Figure 8.13b) is an opioid receptor antagonist of both the μ- and κ-receptors, with a lower affinity for δ-opioid receptors. Naltrexone is used to prevent the problem of alcohol craving once the alcohol abuser has ceased alcohol consumption, and will help to maintain abstinence. The activity of naltrexone is believed to be due to both the parent drug and the 6-β-naltrexol metabolite. Ethanol induces the release of endogenous opioids in specific brain regions as well as interacting with the dopaminergic system. In addition, the endogenous opioids may also inhibit GABA interneurons, releasing dopaminergic neurons from inhibition.[72] Naltrexone exerts its effect as an opioid receptor antagonist, showing a high affinity for μ-opioid receptors and reducing alcohol-induced dopaminergic activity. Some interest has been focused on individuals who have a single nucleotide polymorphism (SNP) in exon 1 (118 A > G) of the μ-opioid receptor gene (*OPRM1*), which results in an asparagine to aspartate substitution at position 40 of the μ-opioid receptor (Asn40Asp). The binding to β-endorphin will be approximately three times tighter than the most common allelic form of the receptor.[73] Subjects with this SNP show lower relapse rates compared with those carrying the AA genotype. A variation in the *DRD4* gene, which encodes the dopamine D_4 receptor, may predict better response to naltrexone.

Acamprosate is the calcium salt of *N*-acetylhomotaurine, with similarities to many amino acids, most notably glutamate, GABA, aspartate, glycine and taurine (Figure 8.13c). Its mode of action is possibly *via* the glutamatergic system, by acting as a weak partial co-agonist at the polyamine site on the NMDA receptor complex,[74] low concentrations enhancing activity when receptor activity is low and at high concentrations inhibiting activation when the receptor activity is high. Acamprosate may also have an inhibitory action on mGlu5 receptors.[75] Such results indicate that interaction of the drug with these receptors will be dependent on the circulating concentration of acamprosate as well as the specific subtypes of receptor present. It may also promote the release of taurine in specific brain regions, this sulfated amino acid possibly showing some neuro-modulating effects.[76] Overall, each of these actions produced by acamprosate may contribute to a reduced hyperexcitability in the brain.

Acamprosate and naltrexone are the two main drugs in clinical use for the treatment of alcohol abusers once abstinence has been achieved. Both of these drugs help to prevent alcohol craving. It has recently been reported that the typing of each patient according to various psychiatric and psychological factors dependence will predict which is the better drug to gain a therapeutic response. Figure 8.14 shows the Lesch typology,[77] which has attempted to integrate biological, social and psychological factors of the alcohol abusers into one classification. Four subtypes of alcohol abusers are described. Type 1 are patients who exhibit extreme alcohol withdrawal symptoms (*e.g.* seizures) and who use alcohol to weaken detoxification symptoms. Type 2 are alcohol abusers who use alcohol as self-medication to control anxiolytic effects. Type 3 are alcohol abusers where there are underlying affective disorders. Type 4 patients show pre-morbid cerebral effects, behavioural disorders and a high

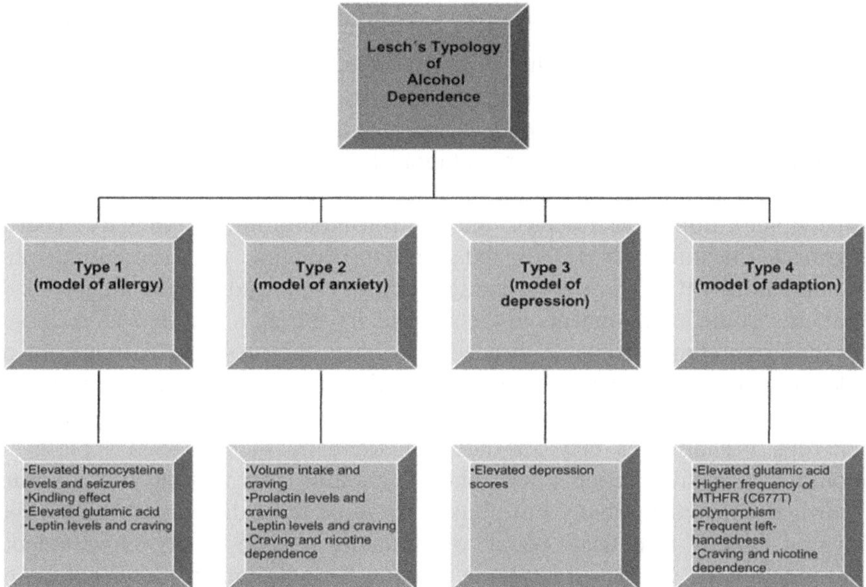

Figure 8.14 Lesch typology.
(Reproduced from Lesch and Walter[77] with permission from OUP).

social burden. As can be seen, acamprosate shows a therapeutic effect particularly in Type 1 and possibly in Type 2 patients. In contrast, Type 3 and 4 patients respond positively to naltrexone.

In other studies, certain polymorphisms have been identified which may enhance the therapeutic response; for example, a genetic variant rs1799971, a single-nucleotide polymorphism in exon 1 of the *OPRM1* gene, which encodes the μ-opioid receptor, may predict better treatment response to opioid receptor antagonists such as naltrexone; variation in the *DRD4* gene, which encodes the dopamine D_4 receptor, may predict better response to naltrexone and olanzapine; a polymorphism in the serotonin transporter gene *SLC6A4* promoter region appears to be related to differential treatment response to sertraline, depending on the subject's age of onset of alcoholism, while a genetic variation in *SLC6A4* may also be associated with better treatment response to ondansetron, a 5-HT_3 receptor antagonist. Clearly more research is needed to understand the mechanisms of these pharmacogenetic interactions and their moderators in order to translate them into clinical practice.

8.8.1 Thiamine

The importance of thiamine supplementation has already been discussed. Reversing thiamine deficiency could potentially prevent damage in specific brain regions and may lead to the improvement of cognitive dysfunction.

8.8.2 Agonists and Antagonists of Neurotransmitter Systems

A number of compounds have been studied in small groups of subjects with alcohol abuse and dependence to assess their efficacy. These include selective 5-HT reuptake inhibitors (fluoxetine), 5-HT(1) receptor agonists (buspirone), 5-HT(2) receptor antagonists (ritanserin), 5-HT(3) receptor antagonists (ondansetron), dopamine receptor antagonists (aripiprazole and quetiapine), dopamine receptor agonists (bromocriptine), GABA(B) receptor agonists (baclofen), NMDAR antagonists (ketamine and memantine) and cannabinoid-1 [CB(1)] receptor antagonists. In early studies of small groups of patients, some improvements in the subjects have been noted. However, larger randomized clinical trials are needed to confirm their role in the treatment of alcohol-dependent subjects. In addition, multi-action therapeutic agents will be needed which will specifically target receptors known to be involved in alcohol craving. Perhaps this may alleviate the desire by such affected subjects to continue excessive amounts of alcohol. It remains clear that alcohol is a drug which is enjoyed by many but abused by some individuals, such that it has proved difficult to educate people on its many adverse effects. Governments throughout the world will continue to strive for solutions to the problems of alcohol abuse, which clearly have enormous social and medical consequences.

References

1. E. B. Rimm, E. B. P. Williams, K. Fosher, M. Criqui and M. J. Stampfer, *Br. Med. J.*, 1999, **319**, 1523.
2. I. Gigleux, J. Gagnon, A. St. Pierre, B. Cantin, G. R. Dagenais, F. Meyer, J. P. Despres and B. Lamarche, *Br. J. Nutr.*, 2006, **136**, 3027.
3. C. Power, B. Rodgers and S. Hope, *Lancet*, 1998, **352**, 877.
4. J. A. Luchsinger, M. X. Tang, M. Siddiqui, S. Shea and R. J. Mayeux, *J. Am. Geriatr. Soc.*, 2004, **52**, 540.
5. R. T. Gun, N. Pratt, P. Ryan, I. Gordon and D. Roder, *Aust. N. Z. J. Public Health*, 2006, **30**, 318.
6. E. V. Sullivan and N. M. Zahr, *Exp. Neurol.*, 2008, **213**, 10.
7. K. F. Mann, K. Ackermann, B. Croissan, G. Mundle and A. Diehl, *Alcohol: Clin. Exp. Res.*, 2005, **29**, 896.
8. S. Campanella, P. Peigneux, G. Petit, F. Lallemand, M. Saeremans, *et al.*, 2013, **8**, e2260.
9. R. P. Vetreno, J. M. Hall and L. M. Savage, *Neurobiol. Learn. Mem.*, 2011, **96**, 596.
10. G. F. Koob, A. J. Roberts, G. Schulteis, L. H. Parsons *et al.*, *Alcohol: Clin. Exp. Res.*, 1998, **22**, 3.
11. N. M. Zahr, K. L. Kaufman and C. G. Harper, *Nat. Rev. Neurol.*, 2011, **7**, 284.
12. C. G. Harper, J. J. Kril and J. Daly, *Br. Med. J.*, 1987, **294**, 534.
13. E. V. Sullivan, A. Deshmukh and J. E. Desmond, *Alcohol: Clin. Exp. Res.*, 1998, **22**, 63A.

14. A. Pfefferbaum and E. V. Sullivan, *Neuropsychopharmacology*, 2005, **30**, 423.
15. C. L. Dirksen, J. A. Howard, A. Cronin-Golomb and M. Oscar-Berman, *Neuropsychiat. Dis. Treat.*, 2006, **2**, 327.
16. C. G. Harper and I. Matsumoto, *Curr. Opin. Pharmacol.*, 2005, **5**, 73.
17. A. Pfefferbaum, E. V. Sullivan, D. H. Mathalon and K. O. Lim, *Alcohol: Clin. Exp. Res.*, 1997, **21**, 521.
18. D. A. Gansler, G. J. Harris, M. Oscar-Berman, C. Streeter, R. F. Lewis, I. Ahmed, *et al.*, *J. Stud. Alcohol*, 2000, **61**, 32.
19. A. C. Chen, B. Porjesz, M. Rangaswamy, C. Kamarajan, Y. Tang, K. A. Jones, *et al.*, *Alcohol: Clin. Exp. Res.*, 2007, **31**, 156.
20. E. J. Devor and C. R. Cloninger, *Annu. Rev. Genet.*, 1989, **23**, 19.
21. D. M. Dick and L. J. Bierut, *Curr. Psychiat. Rep.*, 2006, **8**, 151.
22. O. M. Lesch, J. Kefer, S. Lentner, R. Mader, B. Marx, *et al.*, *Psychopathology*, 1990, **23**, 88.
23. S. Pombo and O. M. Lesch, *Alcohol Alcohol.*, 2009, **44**, 46.
24. R. J. Ward, W. Kest, P. Bruyer, F. Lallemand and P. De Witte, *Alcohol Alcohol.*, 2001, **36**, 39.
25. R. J. Ward and Ch. Coutelle, *Arch. Womens Ment. Health*, 2003, **6**, 231.
26. D. I. N. Sherman, R. J. Ward and T. J. Peters, in *Enzymology and Molecular Biology of Carbonyl Metabolism*, ed. H. Weiner, R. S. Holmes and B. Wermouth, Plenum, New York, 1994, p. 595.
27. A. Yoshida, V. Dave, R. J. Ward and T. J. Peters, *Ann. Hum. Genet.*, 1989, **53**, 1.
28. J. J. Miguel-Hidalgo, *Curr. Drug Abuse Rev.*, 2009, **2**, 72.
29. V. M. Karpyak, J. R. Geske, C. L. Colby, D. A. Mrazek and J. M. Bienacka, *Addict. Biol.*, 2012, **17**, 798.
30. J. P. Ridge, A. M. Ho, D. J. Innes and P. R. Dodd, *Ann. N. Y. Acad. Sci.*, 2008, **1139**, 10.
31. J. P. Ridge, A. M. Ho and P. R. Dodd, *Alcohol Alcohol.*, 2009, **44**, 594.
32. T. Yakovleva, I. Bazov, H. Watanabe, K. F. Hauser and G. Bakalkin, *Brain Behav. Immun.*, 2011, **25** (suppl. 1), S29.
33. K. Suk, *Curr. Neurovasc. Res.*, 2007, **4**, 131.
34. Y. O. Nunez and R. D. Mayfield, *Front. Genet.*, 2012, **3**, 1.
35. R. C. Miranda, A. Z. Pietrzykowski, Y. Tang and P. Sathyan, *et al*, *Alcohol: Clin. Exp. Res.*, 2010, **34**, 575.
36. I. Matsumato, *Alcohol Alcohol.*, 2009, **44**, 171.
37. K. Alexander-Kaufman, G. James, D. Sheedy, *et al.*, *Mol. Psychiat.*, 2006, **11**, 56.
38. K. Alexander-Kaufman, S. Cordwell, C. Harper, *et al.*, *Clin. Appl.*, 2007, **1**, 62.
39. K. M. Adams, S. Gilman, R. A. Koeppe, *et al.*, *Alcohol: Clin. Exp. Res.*, 1993, **17**, 205.
40. M. H. Dao-Castellana, Y. Samson, F. Legault, *et al.*, *Psychol. Med.*, 1998, **28**, 1039.

41. R. Z. Goldstein, A. C. Leskovjan, A. L. Hoff, *et al.*, *Neuropsychologia*, 2004, **42**, 1447.
42. I. Suarez, G. Bodega and B. Fernandez, *Neurochem. Int.*, 2002, **41**, 123.
43. N. M. Zahr, R. L. Bell, H. N. Ringham, E. V. Sullivan, *et al.*, *Pharmacol. Biochem. Behav.*, 2011, **99**, 428.
44. L. Hipolito, L. Marti-Prats, M. J. Sanchez-Catalan, A. Lolache and L. Granero, *Neurochem. Int.*, 2011, **59**, 559.
45. C. J. McClain, S. Barve, I. Deaciuc, M. Kugelmas and D. Hill, *Semin. Liver Dis.*, 1999, **19**, 205.
46. R. Yirmiya and I. Goshen, *Brain Behav. Immun.*, 2011, **25**, 1008.
47. L. Qin and F. T. Crew, *J. Neuroinflammation*, 2012, **18**, 130.
48. S. Alfonso-Loeches and C. Guerri, *Crit. Rev. Clin. Lab. Sci.*, 2011, **48**, 19.
49. R. J. Ward, Y. Zhang, R. R. Crichton, B. Biret, *et al.*, *FEBS Lett.*, 1996, **389**, 119.
50. A. Okvist, S. Johansson, A. Kuzmin, I. Bazov, *et al.*, *PLoS One*, 2007, **2**, e930.
51. J. He and F. T. Crews, *Exp. Neurol.*, 2008, **210**, 349.
52. J. J. Miguel-Hidalgo, *Curr. Drug Abuse Rev.*, 2009, **2**, 72.
53. P. M. Abdul Muneer, S. Alikunju, A. M. Szlachetka, A. Mercer, *et al.*, *Int. J. Physiol. Pathophysiol. Pharmacol.*, 2011, **3**, 48.
54. M. Tomás, E. Fornas, L. Megías, J. M. Durán, *et al.*, *J. Neurochem.*, 2002, **83**, 601.
55. N. A. Floreani, T. J. Rump, P. M. Abdul Muneer, S. Alikunju, *et al.*, *J. Neuroimmune Pharmacol.*, 2010, **5**, 533.
56. S. L. Vallés, A. M. Blanco, M. Pascual and C. Guerri, *Brain Pathol.*, 2004, **14**, 365.
57. S. S. Greenberg, O. Jie, X. Zhao, J. F. Wang and T. D. Giles, *Alcohol: Clin. Exp. Res.*, 1998, **5**(suppl), 260S.
58. S. Fernandez-Lizarbe, M. Pascual and C. Guerri, *J. Immunol.*, 2009, **183**, 4733.
59. S. Alfonso-Loeches, M. Pascual-Lucas, A. M. Blanco, I. Sanchez-Vera and C. Guerri, *J. Neurosci.*, 2010, **30**, 8285.
60. S. Alfonso-Loeches, M. Pascual, U. Gomez-Pinedo, M. Pascual-Lucas, J. Renau-Piqueras and C. Guerri, *Glia*, 2012, **60**, 948.
61. S. Sriram, *J. Neuroimmunol.*, 2011, **239**, 13.
62. W. B. Coleman and C. C. Cunningham, *Biochim. Biophys. Acta*, 1991, **1058**, 178.
63. S. Manzo-Avalos and A. Saavedra-Molina, *Int. J. Environ. Res. Public Health*, 2010, **27**, 4281.
64. R. J. Ward, C. Abiaka and T. J. Peters, *J. Nephrol.*, 1994, **7**, 89.
65. L. A. Howard, S. Miksys, E. Hoffmann, D. Mash and R. F. Tyndale, *Br. J. Pharmacol.*, 2003, **138**, 1376.
66. F. Fadda and Z. L. Rossetti, *Prog. Neurobiol.*, 1998, **56**, 385.
67. X. S. Deng and R. A. Deitrich, *Curr. Clin. Pharmacol.*, 2007, **2**, 145.
68. R. J. Ward, F. Lallemand and P. De Witte, *Alcohol Res. Health*, 2004, **9**, 273.

69. T. J. Rump, P. M. Abdul Muneer, A. M. Szlachetka, A. Lamb, *et al.*, *Free Radical Biol. Med.*, 2010, **49**, 1494.
70. J. Nagy, *Curr. Neuropharmacol.*, 2008, **6**, 39.
71. K. S. Barth and R. J. Malcolm, *CNS Neurol. Disord.: Drug Targets*, 2010, **9**, 5.
72. L. A. Ray, P. F. Chin and K. Miotto, *CNS Neurol. Disord.: Drug Targets*, 2010, **9**, 13.
73. C. Bond, K. S. Laforge, M. Tian, D. Melia, S. Zhang, L. Borg, J. Gomg, J. Schluger, J. A. Strong, S. M. Leal, J. A. Tischfield, M. J. Kreek and L. Yu, *Proc. Natl. Acad. Sci. U. S. A.*, 1998, **95**, 9608.
74. K. Naassila, S. Hammoumi, E. Legrand, P. Durbin and M. Daoust, *Alcohol: Clin. Exp. Res.*, 1998, **22**, 802.
75. P. De Witte, J. Littleton, P. Parot and G. Koob, *CNS Drugs*, 2005, **19**, 517.
76. A. Danchour and P. De Witte, *Prog. Neurobiol.*, 2005, **60**, 343.
77. O. M. Lesch and H. Walter, *Alcohol Alcohol.*, 1996, **31**(Suppl 1), 59.
78. H. P. Rang, M. M. Dale and J. M. Ritter, *Pharmacology*, Churchill Livingstone, Edinburgh, 4th edn, 2001, p. 2.
79. F. Ducci and D. Goldman, *Addiction*, 2008, **103**, 1414.
80. C. C. Y. Wong, J. Mill and C. Fernandes, *Addiction*, 2011, **106**, 480.
81. M. Naoi, W. Maruyama and G. M. Nagy, *Neurotoxicology*, 2004, **25**, 193.
82. H. J. Wang, S. Zakhari and M. K. Jung, *World J. Gastroenterol.*, 2010, **16**, 1304.

Subject Index

Locators in **bold** refer to figures/tables